KU-779-786

Technology and organization

The impact of technology in organizations is of central concern to managers. It plays a crucial role in job design, the motivation of the individual at work, the successful management of change and the structure of organizations. Earlier literature has intended to fall into two separate camps: either specialized studies of new technologies or mainstream accounts of organizational behaviour in which technology is a peripheral concern.

In this important new MBA text Scarbrough and Corbett adopt a highly integrative approach. Using the three conceptual lenses of *power, meaning*, and *design* they explore fully the many different ways in which technology and organizations interact. Technology can just be a description of the productive hardware. It can also be less a question of specific technologies-in-use and more concerned with the impact of technology upon organizational structures and the power balances in complex organizations. Finally, the human factors of perception and knowledge lend technology and its associated debates a further twist. This book highlights the major debates within these competing perspectives and argues that the flow of knowledge and ideas within and between organizations is ultimately crucial in shaping technologies and organizations alike.

The challenging arguments presented by Scarbrough and Corbett should make essential reading both for experienced managers and students of organization who are no longer content to stick within the confines of many of the current recipe-driven solutions to technological issues.

Harry Scarbrough and **J. Martin Corbett** are lecturers in Industrial Relations and Organizational Behaviour at Warwick Business School.

The Routledge series in analytical management
Series editor: David C. Wilson
University of Warwick

This series is a welcome new resource for advanced undergraduate and post-experience students of management who have lost patience with 'off the shelf' recipes for the complex problems of strategic change. Individual series titles cross-reference with each other in a thoroughly integrated approach to the key ideas and debates in modern management. The series will be essential reading for all those involved with studying and managing the individual, corporate and strategic problems of management change.

Other titles available in the series

A Strategy of Change:
Concepts and controversies in the management of change
David C. Wilson

Strategies of Growth:
Maturity, recovery and internationalization
Peter McKiernan

Forthcoming

Competitiveness and Chaos
Walter Dean and Richard Whipp

Managing Culture
David Gray and Geoffrey Mallory

What is Strategy and Does it Matter?
Richard Whittington

Technology and organization

Power, meaning and design

Harry Scarbrough
and
J. Martin Corbett

London and New York

First published 1992
by Routledge
11 New Fetter Lane, London EC4P 4EE

Simultaneously published in the USA and Canada
by Routledge
a division of Routledge, Chapman and Hall, Inc.
29 West 35th Street, New York, NY 10001

© 1992 Harry Scarbrough and J. Martin Corbett

Typeset in 10/12pt Baskerville
by Leaper & Gard Ltd, Bristol, England
Printed and bound in Great Britain
by Mackays of Chatham PLC, Kent

All rights reserved. No part of this book may be reprinted or
reproduced or utilized in any form or by any electronic,
mechanical, or other means, now known or hereafter
invented, including photocopying and recording, or in any
information storage or retrieval system, without permission in
writing from the publishers.

A catalogue reference for this title is available from British Library

ISBN 0–415–07384–7
0–415–05941–0

Library of Congress Cataloging in Publication Data
Scarbrough, Harry, 1955–
    Technology and organization : power, meaning, and design / by Harry
Scarbrough & J. Martin Corbett.
    (The Routledge series in analytical management)
    ISBN 0–415–07384–7 (HB). – ISBN 0–415–05941–0 (PB)
    1. Technology–Management.   I. Corbett, J. Martin, 1956– .
    II. Title.   III. Series.
    T49.5.S33   1992   91–44793
    658.5′7–dc20   CIP

Contents

Figures

Tables

Foreword

The subject of technology – especially the advent of new technologies – has occupied a substantial part of our knowledge about organizations. Ranging from individual motivation through job design to the management of change, the impact of technology is of central concern to managers and those who seek to analyse the impact of technology. In this book, Scarbrough and Corbett take a highly integrative approach towards the subject of technology using three conceptual lenses (power, meaning and design) to allow a thorough analysis of the competing and complementary issues surrounding technology and organizations.

The book thus fills many gaps in the existing literature by proposing different perspectives within the one volume. Technology can be just a description of the productive hardware, of course. It can also be far less a question of specific technologies-in-use and more concerned with the impact of technology upon organizational structures and the power balances in complex organizations. Finally, the human factors of perception lend technology and its associated debates a further twist. This book highlights the major debates within these competing perspectives and argues that the power of technology to sculpt and shape organizational strategies and structures is paramount. The challenging arguments proposed by Scarbrough and Corbett should make essential reading for both experienced managers and students of organization who are no longer content to stick within the confines of many of the current recipe-driven solutions to technological issues.

David C. Wilson (Series Editor)
University of Warwick, 1991

Chapter 1

Introduction
Knowing the dancer from the dance

How can we know the dancer from the dance?
W.B. Yeats ('Among School Children VIII')

INTRODUCTION

This book has a narrow spine but a wide embrace. In addressing the relationship between technology and organization it takes in an extensive but highly fragmented set of ideas and studies, including topics as diverse as job design, product innovation and technology implementation. Our first task, therefore, is to relate these localized perspectives to the overarching relationship between technology and organization.

The metaphor of the dance, which provides the initial theme here, may seem to be a rather fanciful way of viewing this relationship. However, the images that best evoke current patterns of human–technology interaction – the trance-like state of 'Walkman' wearers, for instance, or the addictive lure of the video game, or even the writer enslaved by the screen of the word-processor PC – seem to demand a richer vocabulary of action than the common-sense notion that we simply 'use' technology.

At the organizational level too, which provides the main focus for this book, the need for such a vocabulary is equally great. Although Information Technology (IT) is but one of the many technologies we will examine, its widespread application since the early 1980s has served to challenge both theoretical and practical understandings of technology. As an information-based technology, IT and its associated expertise comes closest to defying the conventional distinction between social structures and technical systems. Its lightning-fast streams of information and communication seem ready to displace the sluggish flows of bureaucracy and paper on which such distinctions are based. The development of 'expert systems' and of 'hybrid managers', which we touch on in Chapter 7, bears witness to the practical implications of such features of IT.

But, if technology is changing out of all recognition, such change is matched by the evolution of organizational forms. In the 1980s we start to see trends towards the 'disorganization' (Lash and Urry 1987) of industries and corporations. Large-scale bureaucratic organizations fragment and deliquesce into loosely connected networks of production and innovation. There are extensive moves towards the subcontracting of activities, and more flexible forms of interorganizational relationships and alliances emerge. As Rosabeth Kanter puts it, we see the 'Giants learn to dance' (Kanter 1989).

This disorganization of the industrial context creates a challenge for our theoretical understanding of the relationship between technology and organization. The fluidity and interpenetration of technological and organizational forms makes it increasingly difficult to, as our chapter heading puts it, 'know the dancer from the dance'. It no longer seems possible or desirable to answer this question simply by differentiating one 'variable' from the other and looking for the causal relationships between them. Although this kind of polarization is embodied in the, to date, dominant schools of thought in the field − one emphasizing 'technological determinism', the other

Table 1.1 Definitions of technology

Technical		Social
Apparatus	Technique	Organization (Winner 1986)
Physical objects	Hardware-related and skills	Human activities (MacKenzie and Wacjman 1985)
Physical pieces of machinery − instruments of production		Relationships of production (Batstone et al. 1987)

'organizational choice' – its static and sterile opposition of concepts seems increasingly ill equipped to capture the essentially dynamic character of the technology–organization relationship.

Hence, before we come to a more detailed description of the perspectives noted above, we will briefly outline our own distinctive approach. Whereas the determinist and choice positions tend to view technology and organization as the finished articles of structures, systems and machines, our approach involves 'knowing' them as fluid and interlocking *processes*. This is not to discount the importance of organization structures or machine systems. These are real enough to the people who work with or within them, and that 'reality' may have, as we discuss in Chapters 4 and 5, a significant influence on behaviour. However, as we noted of IT and industrial disorganization, the boundaries of structures and systems alike are increasingly seen as simply the provisional outcomes of underlying flows of resources, knowledge and information. This kind of fluidity highlights the extent to which technology and organization evolve and overlap together rather than separately or in opposition to each other.

The implications of this view can be gauged by contrasting it with existing perspectives, outlined on pp. 4–6, which define the relationship in terms of technology determining organization or vice versa. Achieving the kind of theoretical closure to which such views aspire has a number of adverse consequences for our ability to grasp the richness of the technology–organization interaction.

First, it involves divorcing technology from organization. Many writers acknowledge that this is a highly arbitrary, snapshot kind of distinction, in that, bereft of a social context of skills, techniques and knowledge, technology is not only inert, but is literally meaningless. Consequently, as indicated in Table 1.1, technology is often presented as part of a spectrum which ranges from hardware at one extreme to social and organizational structures at the other.

However, the theoretical strategy of making a clear distinction between technology and organization means that the operational definitions used in research have generally tended to embrace what has been termed a 'materialist ontology' (Barley 1990). In other words, technology is defined as *essentially* machinery and hardware (cf. Batstone *et al.* 1987).

Yet, a moment's reflection on, for instance, the development of the computing industry suggests that the material embodiment of technology in hardware may fluctuate wildly according to a variety of social and economic, but not purely 'technical', contingencies.

Thus, the distribution of computing functions between hardware, software and the work of in-house systems staff has varied enormously over the years, thanks very largely to 'non-technical' factors such as marketing or intellectual property (Friedman and Cornford 1989).

Second, these views enshrine a partition between the respective roles of subjective action and the objective forces of science and economics. Thus, the determinists view the technology–organization relationship from the objective standpoint of correlations between 'variables' such as machine-systems and levels of management. The proponents of 'choice', on the other hand, focus on the subjective intentions and responses of interest-groups and decision-makers within the organization.

Technological determinism

This perspective suggests that 'the adoption of a given technical system ... requires the creation and maintenance of a particular set of social conditions as the operating environment of that system, (Winner 1986: 32). In other words, organizations have little choice but to adapt their skills and work organization to the requirements of technology.

This view can be illustrated with the example of the ENIAC computer of 1946. This was the first modern computer. It was 10 feet high and over 100 feet in length. Designed to carry out complex calculations of ballistic trajectories for the US armed forces, the ENIAC contained over 17,000 separate vacuum tubes.

The organizational implications of this early computing technology seemed to be quite straightforward. First, it was a brute of a machine to operate (Stern 1981). Programming it for a new formula involved several days' work laboriously rewiring the circuitry by hand. Second, a whole organizational apparatus had to be developed to service and operate this computer and others like it; a division of labour that ranged from highly qualified scientists through to technicians and data input operators. Special, air-conditioned environments had to be constructed to house the machine, and elaborate organizational arrangements made to communicate and handle its inputs and outputs.

The apparent organizational effects of the ENIAC case can be compared with the more systematic evidence produced by Woodward's (1965) surveys of companies in south-east Essex. In these studies, technology was defined in terms of a typology of production systems that ranged from 'unit' through 'batch' and 'mass' to 'process production'. Woodward found that in relation to the length of the chain of command, and the proportion of indirect to direct labour, each major type of technology had a distinctive impact upon both management structure and the organization of work. On the basis of these findings, Woodward claimed that 'there are prescribed and functional relationships between structure and technical demands' (p. 51), and that 'there was a particular form of organization most appropriate to each technical situation' (p. 72).

Organizational choice

Where technological determinists often take elaborate machine-systems as their model, advocates of organizational choice have generally elected IT as the prototypical technical form. This is, in fact, a supremely 'malleable' technology: miniaturized, portable and programmable. It offers enormous information-processing power yet is readily shaped and reshaped according to the wishes of its users.

In highlighting the processes of change and decision-making surrounding the introduction of IT, this particular school of thought has found that the process of change, and hence the technology itself, is primarily shaped by actors within the organization. Thus: 'Technology has no impact on people or performance in an organisation independent of the purposes of those who would use it, and the responses of those who have to work with it' (Buchanan and Huczynski 1985: 222).

It follows that in any given context, the relationship between technology and organization is largely determined by the managerial intentions and values (Buchanan and Boddy 1983), and organizational politics (Pettigrew 1973) which characterize the organization. Technology simply 'embodies' the intentions and interests of particular groups (Child 1985).

Now, subjective perceptions and actions clearly are important in guiding the development of technology at an organizational level. However, as we discuss later (see pp. 12–22), the degree of autonomy and control which organizations possess in developing technology is by no means reducible to the subjectivity of their managers. More important for our present concerns, though, are the implications of divorcing subjective responses or organizational strategies from an 'objective' technological environment. Apart from the fact that technology itself is shaped by and must perforce enlist such subjective actions, this divorce literally marginalizes the forms of technological knowledge, and the related social and occupational networks, on which organizational choice and decision-making depend.

The importance attached to the diffusion of such knowledge, whether embodied in a technological artefact, or associated with its use, is usefully demonstrated by the relative 'impacts' of the technologies cited as illustrative examples above. Whereas ENIAC had specific effects upon only one organization, IT-related knowledge – or even knowledge about the potential of IT – seems to insinuate itself into just about every area of organizational 'choice'.

The importance of both technological and organizational knowledge is a recurrent theme in this book. Indeed, if the problem of 'knowing the dancer from the dance' has an answer it is probably another question to do with the nature of knowledge and the transformation of subjective experience into objective forms. More specifically, as we outline below, this notion of flows of knowledge helps to make some sense of the processual and mutual interaction – the metaphorical dance – out of which the technology–organization relationship emerges.

TECHNOLOGY AND ORGANIZATION: A PROCESSUAL VIEW

A useful starting point for viewing technology as a process rather than as aggregations of machines and systems (cf. Macdonald 1985, Clark and Staunton 1989) is provided by the linear model of technological innovation outlined in Figure 1.1. In this model – which is discussed in more detail in Chapter 7 – technologies are seen as being generated and diffused by transfers of knowledge and artefacts between three different processes. In broad terms, we can characterize these as, first, an invention process where new ideas and forms of technological knowledge are generated. Second, an exchange process where product-market and labour-market factors influence the design of the technology. And, finally, a use or production process where technologies are applied to particular tasks.

Now, although this model usefully highlights the transformational aspects of technology and the key social processes from which it emerges, in general it clings to a deterministic view of tech-

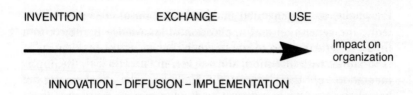

INVENTION   EXCHANGE   USE

Impact on
organization

INNOVATION – DIFFUSION – IMPLEMENTATION

Figure 1.1 A linear model of technological innovation and adoption

nological change. By making questionable assumptions about the operation of market-forces, and the innate superiority of new forms of 'scientific' knowledge, it suggests that organizations have little choice but to apply the most advanced technologies or else go under. It reflects the determinists' sense of technological 'necessity' by welding science, technology, markets and organizations together into an objective and interlocking causal chain. The only remaining uncertainty is how the innovation process itself is initiated, whether by 'technology push' or 'demand pull' (Langrish *et al.* 1972).

However, it is possible to accept the importance of these under-lying activities without accepting the iron logic of linearity and determinism. The processes of invention, use and exchange can better be viewed, for example, not as abstract scientific and economic forces but in terms of subjective actions and loosely coupled forms of social organization. This is not to reduce tech-nology to social relations *per se*, since this would neglect the trans-formational dimension which is one of the most important features of the innovation model. Indeed, it is to accept that, through its inventions and combinations of knowledge, the technology process clearly does transform social relations and actions. It does so not directly through physical constraints, but by the creation of perceived utilities which induce its adoption and use. However, this insidious usefulness of technology, which as we noted earlier is what makes current IT so much more important than the ENIAC type of computer, cannot be defined simply in the abstract, economic terms of 'utility functions'. Rather, the useful aspects of technology are identified, defined, and indeed contested, within specific social forms of organization, where economic criteria may be but one of a number of factors.

Similarly, the proposition that invention, exchange and use are linked together in sequential fashion neglects the possibility for knowledge to be generated in a variety of social contexts. Admittedly, the generation and application of knowledge is subject to a broad societal division of labour, such that we associate scientists in laboratories with invention, and workers in factories with the implementation of that invention. However, the deployment of knowledge-workers in industry, and the increasing pressures (documented in Chapter 3) for 'user involvement' in the technology process, are increasingly standing that proposition on its head.

It is equally unsafe to assume that the technology process is driven by objective or abstract forms of scientific knowledge. Since the ground-breaking work of Kuhn (1962) on 'paradigms' within science, there has been growing recognition of the forms of social organization and consensus which underpin the generation of scientific knowledge. Nor is technology simply the product of scientific invention. In fact, 'where technology does draw on science the nature of that relation is not one of technologists obediently working out the "implications" of a scientific advance.... Technologists *use* science' (MacKenzie and Wacjman 1985: 9).

Scientific knowledge is, therefore, but one element of the forms of knowledge which are assembled within the technology process. At least as important is knowledge of the social processes in which the technology is to be used. Many of the crucial, incremental improvements in process technology, for instance, occur on the shop-floor. Fleck (1987) notes how the development of industrial robotics in the UK was characterized by the many innovations which arose out of the immediate 'context of use'. Thus, even the factory can be seen as a laboratory where different technologies are tested, adapted and modified.

In a similar vein, the deterministic impact of market-forces on the generation and diffusion of technological knowledge may be easily exaggerated. For example, the trading of knowledge may occur through occupational networks and interorganizational relationships, and, as we describe in Chapter 8, the market itself may be shaped by the strategies and knowledges of particular organizations.

Thus, to the extent that the various social processes outlined above overlap and interpenetrate in a variety of ways – with the production process, for instance, being influenced by product-market and labour-market factors – there are a variety of ways in which knowledge and artefacts can be generated and flow between them. Clearly, there is no linear logic to the unfolding of the technology process. Rather, as Figure 1.2 indicates, the flows of know-

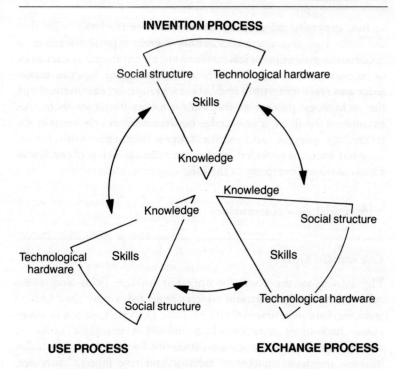

Figure 1.2 The technology process

ledge between different social contexts can more usefully be depicted as cyclical and reciprocal in character.

Indeterminacy in the technology process is reflected in the relationship between technological hardware and knowledge. First, as we note in the later example of the 'mirror problem', artefacts themselves may come to inform the way in which social actors view an organization. But, second, the way in which engineers, say, 'know' the distinction between technology, on the one hand, and the organization and skills of the workforce, on the other, can itself influence the course of social and technological development in the workplace. If they view that distinction in terms of the 'engineering logic' which we outline in Chapter 5 – that is, an objective technological system counterposed to subjective, human 'noise' – this is likely to encourage the emergence of ever more automated forms of process technology. Thus, even 'theoretical' distinctions may be grist to the mill of knowledge.

But, generally, all of these factors reinforce the basic point that the individual organization is neither merely a passive receiver of predetermined technological artefacts nor an autonomous controller of technological change. Rather, in organizing the flows of knowledge and resources within and between groups, organizations shape the technology process at the same time as it shapes them. An example of the flows of knowledge between different elements of the technology process, and of the latter's interaction with organizational forms, is provided by the case example below of the Kwik-Fit car servicing company in the UK.

A processual view of technology

Case example: Kwik-Fit

The growth of the Kwik-Fit operation in the 1970s and 1980s reflected a strategic formula which connected a new approach to servicing cars to a new definition of the market. Unlike the traditional back-street garages which offered a complete range of maintenance and repair services, Kwik-Fit focused on the so-called 'distress purchase' market of exhaust and tyre fitting. Moreover, where the back-street operators were structured around the high levels of technological uncertainty and high skill levels associated with breakdown repairs, Kwik-Fit were able to reduce the level of technological complexity through their focus on a relatively narrow range of simple services.

This segmentation of the market was based on the increasing modularity of the product technology itself, as the design of cars became increasingly oriented to the easy replacement of standardized parts. It had important implications for the design of Kwik-Fit services. First, it allowed Kwik-Fit to promote a readily identifiable, standard range of services through an extensive network of depots, thereby achieving economies of scale. The layout of each depot, with open fitting bays, service and reception areas, gave the customer a sense of control. Similarly, by specializing in simple services – just tyre and exhaust fitting originally – Kwik-Fit were able to enhance customer control by offering 'while you wait' service. Most importantly, service specialization allowed task specialization of the workforce, putting management firmly in control of the service process, and substituting cheap YTS labour for more highly skilled mechanics.

But the advantages of customer transparency and management control also created constraints on the future growth of the business, as control complexity grew with every new depot in the network. Centralized, bureaucratic control would have created enormous central overheads and multiplied administrative inefficiencies at depot level. Recognizing that they had 'an administrative structure that was good for 50 outlets but not for 180', and which was 'beginning to creak', Kwit-Fit management looked to IT for some solutions.

After some abortive contacts with specialist computer suppliers who seemed unable to understand their needs, a chance event led Kwik-Fit management to pursue a relationship with the US fast-food chain, 'Church's Fried Chicken' – the latter having developed a computer system which seemed particularly suitable. Further contacts between the two companies – with one of Church's executives managing a Kwik-Fit depot for a week to understand their operations – led to the development of a similar system for Kwik-Fit. Depot managers, not programmers, helped draw up the specifications of the system.

Computer terminals were installed throughout the Kwik-Fit network. These were titled MATs (management action terminals) for fear of the detrimental effect the word 'computer' might have on staff. They performed all the essential functions of a depot's administration, including quotation and invoice production, the recording of customer and banking transactions, confirmation of stock levels, and the recording of staff working hours for payroll. At the close of trading each evening, the day's transactions are polled and collated by central mainframe computers, and by 7 a.m. the next morning managers at both central and depot level have detailed management and sales information.

The system effectively frees depot managers from paperwork and routine administrative tasks. When allied to a reward structure which links the pay of both fitters and managers to depot sales and costs, it turns Kwik-Fit depots into something akin to small businesses while retaining the economic and informational advantages of the large business.

Source: Gallagher and Scott (1988)

The organization as process

Paradoxically, the disorganization of industry has helped to make the underlying processes which draw groups and resources together more transparent. Business organizations are increasingly seen not as 'islands in the stream' of environmental change, but simply as provisional 'modes of organizing' flows of transactions, knowledge and information (Child 1987a). Thus, like technological artefacts, organization structures can be viewed as the relatively fluid outcome of interactions between processes of invention, production and exchange.

The dynamics of the underlying processes within organizations can be interpreted in many different ways. There are, for instance, vast literatures on the organization as a labour process, or the organization as a decision-making process, which in this context we can only touch on. In broad terms, however, such underlying dynamics can be related to the emergence of forms of control which connect production processes to product-markets and labour-markets. Again, some writers, notably Braverman (1974) and Williamson (1985), have argued that a relatively deterministic logic underpins the emergence of control relations between different groups. However, the relative autonomy of production processes from market-forces and the importance of subjective actions in shaping the development of such processes suggest that in practice the pursuit of control can take many different forms.

A more critical influence, however, may be the reciprocal linkage between organizational control structures and the assembly and invention of forms of organizational knowledge. Thus, as we discuss below, the development of general models of control such as bureaucracy or Fordism both feeds on and promotes a more objective control and analysis of organizational activity. In this context, the 'invention' (Hoskin 1990) of the strategy concept has had important implications for the development of more elaborate forms of control and organizational structures. The development of a strategic level of control allows the adoption of new forms of knowledge to be expressed either through the assimilation of industrial 'recipes' (Child and Smith 1987) or the exercise of 'strategic choice' (Child 1972).

This brief account of the organizing process highlights the importance of knowledge and control in determining organization structures. More importantly, the processual view allows us to

analyse the 'mutuality' (Child 1987a) of technology and organization: that is, the enmeshing of technological and organizational
possibilities and knowledges. Although particular technological and
organizational forms may be shaped initially by different sets of
sectoral contingencies, we find that at the level of design and use
technological and organizational knowledges are combined to
produce distinctive configurations of machines and work organization. For instance, even when technological change is seen as
simply adopting a predetermined 'package' from a supplier in a
different industrial sector, its use will involve combining elements of
technological knowledge with organizational knowledge about tasks
and how they are to be performed. And where an organization's
involvement with technology is more far-reaching, the interaction
between technological and organizational knowledges is likely to
have significant implications for the strategic design of the organization itself as well as for the design of the technology. This kind of
mutuality between technology and organization is expressed in
Figure 1.3.

A variety of terms has been coined to describe different aspects
of the enmeshing of technology and organization; 'orgware' and
'structural repertoires' (Whipp and Clark 1986), for instance, are
terms which we will encounter in later chapters. For the moment,
however, the particular label we attach to this area of mutuality
seems less important than outlining some of the ways in which it can

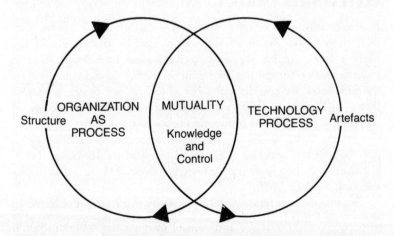

Figure 1.3 The relationship between technology and organization

be interpreted. For example, this sense of an overlapping core of technological and organizational knowledge gives us a different perspective on the relationship between technology and organization structure. Apparent 'correlations' between technology and organization structure look more like 'reflections' of the working through of a common admixture of technological and organizational knowledge. This sense of technological artefacts and organization structures *mirroring* each other may also be expressed in technological design as technologies come to automate or embody organizational rules and controls (Perrow 1972, Woodward 1970). For example, the rules about time-keeping which were a critical aspect of control, and of management–worker dispute, in the early development of the factory system (Marglin 1974) found expression in the development of elaborate 'clocking-in' machines. Similarly, in the development of corporate IT systems, the rules expressing hierarchical access to information have come to be incorporated in the provision of levels of access and 'security', complete with elaborate passwords and gates.

More generally, the relationship between organizational and technological forms can be understood in terms of the patterns of interactions that take place; at different levels of technology; through learning and adaptation over time; and in the relationship between subjective action and objective knowledge.

LEVELS OF TECHNOLOGY

In the continuum between abstract knowledge and concrete artefact, three major levels of technology can be discerned. At the highest level, we have generic technological knowledge such as engineering or computing, and generic models of management. At the next level, we can discern a broad technology design or 'architecture' (Clark *et al.* 1988), such as might be involved in the overall layout of a new plant or production process, for instance. Finally, at the lowest level, we have the implementation of technology as specific items of machinery or 'technical systems' (Berniker 1987), together with the contingent knowledge associated with that implementation (Fleck 1987).

Organizations interact with technologies at a number of levels. In some cases, this interaction may be confined to the lowest level of technology. An off-the-shelf software package, for instance, embodies technological knowledge in a relatively commodified form: knowledge has been translated into a physical product allowing it to be mediated by a market-based exchange process. However, even the purchase of technological artefacts often involves the transfer of some degree of technological knowledge. As the artefact becomes more complex or less 'stabilised' (Pinch and Bijker 1987), higher levels of technological knowledge may need to be acquired. These may involve the hiring of expert consultants, as happens with elaborate Flexible Manufacturing Systems (FMS), for instance. Alternatively, as with techniques such as Just-In-Time or quality circles, it may be a question of exposing in-house management to new forms of cultural knowledge through the mandatory visit to Japan.

Where an organization is seeking to develop technology for long-term strategic purposes, there may be a need to address the highest levels of technological knowledge. This can be described in terms of the nurturing of 'core competences' (Pralahad and Hamel 1990). The latter provide the 'knowledge base' for specific product and process designs, and implementations of technology. Such a strategy is likely to involve recruiting and retaining knowledge-workers who are able to generate competitive and proprietary forms of knowledge. It may even lead, as we note in Chapter 8, to the formation of interfirm alliances which are capable of providing high level flows of knowledge and expert personnel.

LEARNING AND ADAPTATION

Although we cannot talk of physical artefacts having a history, one of the advantages of defining technology as a process is that it does open up this historical dimension. Although some (e.g. Piore and Sabel 1984) would argue that IT is associated with a significant discontinuity or divide in industrial development, it is far from clear that this technology alone has displaced the kind of deep-rooted ideas and perceptions of technology which are discussed in Chapter 4.

Nor does this invalidate the more general point, that there is an important time dimension to the interweaving of technology and organization. As social groups engage in the use process for a particular technology, there is a process of learning and adaptation. Skills develop, knowledges accumulate and means clarify ends. The end result may be that the forms of knowledge and the dependencies associated with the use of a particular technology may create their own implications for social organization, quite independent of the original intentions of the technology's designers. A classic example is provided by the introduction of the snowmobile into the Skolt Lapp community outlined below.

Case example: the Skolt Laplanders and the snowmobile

Introduced in the early 1960s, the snowmobile was adopted by the Skolt Lapp people to replace reindeer sleds as a means of transportation, thereby allowing easier access to trading posts, to better health care, and to a more varied diet and recreation. But within a few years the technology had made a profound impact on the Skolt Lapp community. The older men, who had previously enjoyed high status and authority within the community, lacked the strength to ride the snowmobiles. As a result, their status declined relative to the younger, stronger, men. This status shift was increased by the decline in importance of the elders' knowledge and wisdom concerning the care and use of reindeer herds (which themselves declined owing to a rapid drop in calf births occasioned by the frightening noise of the snowmobiles' engines).

Most important of all, the Skolt Laplanders quickly found themselves dependent on suppliers of imported petroleum and spare parts for the snowmobiles. Thus, an apparently neutral technology brought about significant and largely irreversible changes to a community.

Source: Pelto (1973)

The Skolt case usefully evokes the way in which, over time, technology-related knowledge and skills may become interwoven into the fabric of a social or organizational hierarchy. One possible consequence, which is explored in Chapter 8, is the fusion of technological and organizational goals and commitments.

SUBJECTIVE AND OBJECTIVE FORMS OF KNOWLEDGE

As we indicated earlier, the metaphor of the dance allows us to avoid rigid distinctions between subjective action and objective forms of knowledge. Rather than viewing objective knowledge in terms of an approximation to an external 'reality', and hence a deterministic 'constraint', this metaphor encourages us to see such knowledge as a particular expression of subjective experience. This applies even in a scientific context, as the example of the 'dancing Wu Li Masters' indicates below. And at the organizational level this insight gives us some clues as to the way in which more objective forms of knowledge can both shape and be shaped by subjective choice and experience.

The dancing Wu Li Masters

The issue can be illustrated with reference to a parallel problem in the physical sciences. This is the emergence, in this century, of the 'new physics' and its attempt to transcend Newtonian physics in the study of subatomic matter. New fields such as quantum mechanics arose out of a growing recognition that the 'disorganized' nature of the universe precluded the possibility of a complete, predictive theory of external physical reality. In particular, incompatible experimental findings on the physical nature of light – which seemed in some observations to behave like a particle, and in others like a wave – led to the realization that the concepts which were being tested were not even approximations to objective physical entities, but rather reflected the subjective senses of human scientists. As Henry Pierce Stapp, a physicist at the Lawrence Berkeley Laboratory, puts it: 'The theoretical structure did not extend down and anchor itself on fundamental microscopic space–time realities. Instead it turned back and anchored itself in the concrete sense realities that form the basis of social life' (Stapp 1972).

Theory was reorganized around this key insight: knowledge based on observations of external reality is not simply a reflection of objective, physical forces but is simultaneously a reflection of subjective experience.

Concepts emerged which sought to express this unity of subjective and objective knowledge. One phrase coined to describe the physicists whose work embodies such concepts, and which artfully prefigures our present theme, talks of 'the dancing Wu Li Masters' (Zukav 1979). Zukav describes the proponents of the new physics in the following terms: 'Most people believe that physicists are explaining the world. Some physicists even believe that, but the Wu Li Masters know that they are only dancing with it' (Zukav 1979: 35).

First, as we describe in more detail in Chapter 4, it is important to recognize the extent to which perceptions and 'truths' which are generated in knowing about an external 'reality' also provide the means of shaping that reality. Thus at a historical level, the development of technology and organization alike has reflected the increasing pervasiveness of objective forms of knowledge through the 'rationalization' of society (Weber 1968). And in this context, technology has long provided a model for the pursuit of objectively efficient forms of organization. The word organization itself derives from the Greek 'organon' meaning 'tool', and the aim of much organizational design has been to attain the orderly and predictable qualities associated with mechanical forms of technology.

The 'war machine' of the military provided an early model of mechanistic organization. In particular, Frederick the Great of Prussia developed elaborate forms of organization for his army which were inspired in part by his fascination with the workings of automated toys. As Morgan (1986) notes, his aim of reducing his soldiers to automatons foreshadowed the development of classical models of bureaucracy based on Prussian military organization.

More recently, the development of management and organization theory has similarly derived a good deal of its legitimacy from claims to rationality and objectivity. The centrality of such claims is reflected in the extensive use of technological metaphors in organization theory: from the analogy with mainframe computer 'programs' and 'routines' in the 1950s (March and Simon 1958) to ideas derived from the spread of IT 'networks' in the 1980s.

But the relationship between technological forms and organizational forms is not one way. Very often the application of technological knowledge depends on the prior emergence of organizational methods and controls aimed at translating voluntary and subjective action into the more ordered and objective texture of programmed rules and procedures. The 'scientific methods' of Taylorism, and, in the more recent period, of operations research and systems analysis may differ in their end results, but all are equally concerned with translating a complex social reality into an objective analysis of work flow and operations. In fact, the purely organizational 'innovations' arising from such methods may be at least as significant as the gains produced by more nearly technological means.

Such distinctions between organizational and technical innovations may be short-lived, however, as once the subjective skills of particular tasks have been centralized and codified into objective forms of control, the way lies open for the application of high-level technological knowledge. As Zuboff notes:

> The progress of automation has been the result of a transfer of knowledge ... knowledge was first transferred from one quality of knowing to another – from knowing that was sentient, embedded, and experience-based to knowing that was explicit and thus subject to rational analysis and perpetual reformulation.
> (Zuboff 1988: 56)

The increasing spread of automation and the centralization of task knowledge need to be placed in context, however. The pursuit of completely objective forms of process control is best seen as an ideal rather than a practical objective of technological design. Subjective behaviour and responses continue to be important even in highly automated plant, as studies of productivity (Gallie 1978) and operator error (Perrow 1983) in such settings demonstrate.

The persistence of subjectivity within organizations is partly to do with the sheer breadth and depth of human skill and creativity. Even the most rigorous forms of de-skilling can do little to eliminate the so-called 'tacit skills' (Manwaring and Wood 1985) of the workforce. Similarly, we can cite the relative failure of 'export systems' (see Chapter 7) to incorporate many of even the simplest forms of human knowledge and language within their rule-bound programs.

Again, organizations and markets are not static. The increasing tendency for organizations to privilege innovation over efficiency (Clark and Starkey 1988) when confronted by a rapidly changing market context, places a particular stress upon the human qualities of flexibility and creativity which even the 'impersonal machinery' (Thompson 1990) of IT hardware cannot yet provide. As a result, technological change in production may have the effect of reducing the subjective discretion enjoyed by one group of workers only at the cost of enhancing the skills and bargaining power of yet another group.

But the most important constraint on the dominance of objective forms of knowledge and control is the extent to which the latter actually derive from and enlist the human and subjective elements of the organization. Technology and related forms of knowledge often take on an objective character. They appear to us as an alien force embodying the authority of objectivity and technical necessity. A simple example of that authority occurs in the translation of subjective thoughts and ideas into published form. Compare this word-processed and printed text with the original hand-written version:

Compare this printed & computer word-processed
version text with the original hand-written
version.

Yet, as we note in Chapter 7, objective forms of knowledge and their physical embodiment in machine-systems are actually the product of the organized but still subjective creativity and co-operation of groups such as scientists, engineers and systems analysts. Moreover, the use of technology involves enlisting the subjectivity of an ever-wider range of groups within the organization. Indeed, as IT applications spread throughout the organizational hierarchy it is not just the 'involvement' of shop-floor employees which is being sought, but even that of the senior ranks of management themselves. The latter group, in fact, provides the classic example of the importance of enlisting subjectivity, as outlined in the following case example of Management Information Systems.

Case example: Management Information Systems

The development of Management Information Systems (MIS) is widely touted as a means of providing more rational and objective forms of decision-making and control within organizations. The design of such systems, however, is often based upon a highly objectified and unreal set of assumptions about managerial work (Martin 1986). This neglects the manager's interest in 'live' (i.e. political and topical) information (Mintzberg 1973) – in favour of the objectively determined need for copious amounts of quantitative data. But this is not the only reason why many MIS have failed to improve managerial performance. A more fundamental reason may be the threat that MIS are seen as posing to the discretion and self-image of senior managers. Not only are senior managers reluctant to use VDUs because of their association with clerical work, but the use of technology may undermine the personal qualities of 'charisma' and leadership on which their effectiveness is based (March and Sproull 1990).

The persistence of this kind of subjective–objective tension has important implications for our view of technology. At all levels, it appears – from the generation of technological knowledge through to the use of machinery in production – technology interacts with the subjectivity of those who develop or use it. Weick goes so far as to suggest that new technologies be described as 'parallel technologies' because they involve 'a technology in the head and a technology on the floor' (Weick 1990: 17).

ORGANIZATIONAL AUTONOMY AND TECHNOLOGICAL DEVELOPMENT

In describing various aspects of the coupling of technology and organization we are also reframing the debate about the degree of choice or determinism that exists between them. In this context, it is useful to distinguish between what we can term the 'passive' and the 'active' forms of choice which organizations are able to exercise in relation to technology. In a passive or formal sense, the importance of the subjective perceptions of managers and the relative insulation of organizational hierarchies from market forces suggests a great degree of organizational choice. Certainly, at the level of specific systems, organizations are often presented with an almost infinite number of technological permutations in developing new products or production processes.

However, it is questionable whether choice over the purchase of technological artefacts alone is an adequate way of either theorizing or exercising organizational autonomy. If we relate it to the knowledge embodied in or associated with the use of a technology, it becomes clear that even the purchasing of a technological package actually creates constraints on choice. For instance, purchasing the package involves acquiescing in the pre-emptive 'choices' and knowledges of the supplier which are embodied in that package. Given the predominance of a few large technology suppliers in certain sectors – IBM in financial services, for example – this degree of constraint may be significant. As Ramsay and Beirne note: 'much new technology ... is itself produced by a few firms and sold as a package to others, seriously limiting the scope for local management or employee manoeuvre' (Ramsay and Bierne 1992: 4).

It seems that the exercise of active choice over technology involves some degree of control – and not simply in a formal or hierarchical sense – over the flows of knowledge and artefacts out of which technologies emerge. In other words, autonomy and control is achieved by organizing within the technology process and not outside it. The deployment of knowledge-workers, the exchange of knowledge with suppliers, and the translation of subjective experience into objective systems are all elements of such activity. Indeed, it is precisely to the extent that management seek to translate 'choices' into the use of machinery and techniques on the shop-floor that we are able to speak of a technology process. The deals that have to be made, the ringing appeals to technological 'necessity', the knowledges and understandings deployed in 'choosing' and 'using' the technology are all indicative of mutual interchange with a process that extends beyond formal organizational boundaries.

CIRCUITS OF POWER, MEANING AND DESIGN

Although, as we have described, engagement in the technology process involves a diverse set of groups and knowledges ranged both inside and outside the organization itself, there is no predetermined logic to the way in which that process unfolds. Neither scientific advance nor market forces *per se* drive the direction or outcomes of the process. It remains to this final section, therefore, to provide a brief, introductory account of the forces that shape the technology process at its nexus with the individual organization.

Our earlier analysis of technology viewed it in terms of flows of knowledge and artefacts generated and diffused through invention, use and exchange processes. To the extent that such flows – for instance, at the level of technical knowledge or general models of management – are shaped by occupational or sectoral networks it seems that the individual organization itself is but one site for the shaping of the technology that it applies. Indeed, far from

containing or controlling the technology process, the formal boundaries and managerial hierarchies of organization may themselves be
restructured by it. To cite a couple of industrial sectors which will be
touched on in later chapters: both the history of the motor industry
(Abernathy 1978) and more current events in the UK financial
services sector demonstrate the extent to which technologies and
organizations evolve together through the generation and transmission of strategic and technological knowledge at a sectoral level.
Thus, development in both industries has depended upon the
formation of technological – that is, physical, social and cognitive –
infrastructures linking individual firms into networks of components
suppliers and payment clearing systems respectively.

But if the technology–organization relationship cannot be
explained at the organizational level alone, neither is it satisfactory
to view it as simply the product of broader social and economic institutions. While Braverman (1974) suggests that capitalism leads to
technology being applied to the de-skilling and control of labour,
and Cockburn (1985) views technology as conditioned by the patri-
archal nature of society, such blanket answers tend to homogenize
both technologies and organizations. They do little to convey the
uncertainties and interactions of the technology process, nor do they
account for the role played by groups and individuals in resisting or
reshaping the fundamental institutions of society. Indeed, on
occasion the transformational power of technological knowledge
may escape the intentions of the powerful and undermine, and not
simply reproduce, existing social and economic structures.

There is a great wealth of research evidence in subsequent
chapters which explores such unpredictable and diverse outcomes
within the technology–organization relationship. Rather than privi-
leging either the societal or the organizational level of analysis in
reviewing such evidence, we have sought to highlight those connections between the individual organization and the broader industrial
or social context that shape the emergence of the technology
process. The notion of the 'circuit' has been chosen to describe such
connections, both as a means of organizing the extensive empirical
material and because it underlines the extent to which they both
reflect broader patterns of industrial development and yet are also
shaped by the actions and ideas of groups and individuals at organizational level.

For example, one way of viewing the interrelationship of technology and organization is in terms of the assembly of different

elements of knowledge and information: individuals and groups combining knowledges of invention, use and exchange to develop a product or process innovation. In general, assembling such a diverse range of knowledges, within what we will term a circuit of 'design', involves a wide variety of expert groups. However, a simple illustrative example of such activity is provided by the work of Thomas Edison. Edison was one of the last 'system-builders', able to understand and manipulate all of the necessary knowledges in his own right. His invention of the light bulb illustrates the range of knowledge and information – that is, not only scientific knowledge but also knowledge of the market and of the context for use – that may need to be assembled in the design of technology.

Edison and the light bulb

The invention of the light bulb was actually part of a much bigger design project which encompassed a complete system for the generation and distribution of electricity. And here Edison's masterstroke lay not in producing an operational light bulb – that was relatively easy – but in drawing on his knowledge of the whole generating and distribution system to identify the one feature of the light bulb that was vital to commercial success. Edison's critical contribution lay in recognizing that the large-scale use of the light bulb depended on developing a high-resistance filament: existing low-resistance filaments would have required such large copper conductors for wiring circuits as to make the system 'absolutely uncommercial' in cost terms.

Source: Hughes (1987)

However, as we noted earlier in the discussion of subjective–objective forms of knowledge, it is not sufficient to view technology simply as an objective force. It also derives from and enlists subjective experience, and to that extent is influenced by the ideas and meanings which make sense of that experience. Consequently, the design and application of technology within organizations is also influenced by the perceptions and meanings that are attached to that development. Within this 'circuit of meaning', powerful groups may draw on widely held ideas or ideologies to influence the development of technology by shaping the way in which it is understood, alternately mystifying it through technical jargon to retain control (Pettigrew 1973) or relabelling it to jettison negative connotations. Thus, management at Ford of Europe in the 1980s took to referring to their increasing numbers of industrial robots as UTDs (Universal Transfer Devices) to pre-empt possible worker anxieties about job losses.

But it is not only the meanings and ideologies generated by organizational interests which influence the development of technology. The perceptions of external reference groups, and particularly customers, may also play an important role. Thus, Pinch and Bijker (1987) note how, in the late nineteenth-century development of the bicycle, particular innovations were often ultimately successful not in the original terms of their designers but in ways

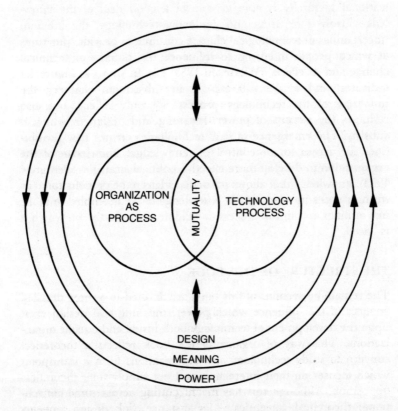

Figure 1.4 Circuits of power, meaning and design

that had meaning for potential customers. Dunlop's air tyre, for example, was developed as a solution to vibration problems, but this aim was not important to the existing constituency of cyclists. Indeed, it was actively derided as the 'sausage tyre' by some. It did finally achieve acclaim, however, when it solved the more meaningful problem (to customers) of winning races.

Finally, the implications of the control of knowledge and the manipulation of meaning highlight the influence of power relations on the technology process. Again, power at an organizational level involves, in part, the active exploitation of a wider circuit of external institutional resources. Thus, the considerable power which management exercise over technology, for instance, is linked to their relations with external constituencies of owners, technology suppliers and providers of capital. However, even when the organizational hierarchy is able to control a good deal of the knowledge involved in using and deploying technology, the inherent uncertainties of technological change continue to provide 'junctures at which people intervene to influence the process of technical change and its effects' (Wilkinson 1983: 92). In short, as Figure 1.4 indicates, and as we will explore in subsequent chapters, the unfolding of the technology process not only reflects but also reshapes the circuits of power, meaning and design in which it subsists. The emergence of new technologies creates new possibilities for access to, or control of, knowledge: knowledge of the organization to develop more effective political strategies (Pettigrew 1973); knowledge that allows particular ideas to be demonstrated as truth or necessity; and, finally, knowledge to be assembled to make new utilities and in the process to validate the expertise on which it is based.

THE STRUCTURE OF THIS BOOK

The remaining sections of this book are devoted to a more detailed analysis of the influence which power, meaning and design exert upon the development of technology both inside and outside organizations. This way of organizing the book reflects a theoretical concern to study technology and organizations from a standpoint which focuses on their interaction and not simply their theoretical opposition. Although this has meant cutting across some conventional theoretical categories – for instance, 'job design' appears under different guises in several different chapters – careful consultation of the index allows such topics to be reassembled and reviewed.

More substantively, the headings around which we have chosen to structure the book could be seen as producing arbitrary divisions between actions and perceptions which in the 'real world' overlap and interact. This is certainly true: the focus on 'knowledge' in the design section clearly has resonances with the study of 'expertise' in the power section or 'ideology' in the meaning section. However, we see these distinctions as at least serving a useful heuristic function, and in some sense they may even help to point up the extent to which the same empirical 'fact' can be understood from a range of sometimes complementary, sometimes conflicting, theoretical perspectives.

Part I

Power

People are trouble but machines obey

Headline in *The Engineer*, 14 September 1978

One consequence of the tendency to equate technology with hardware is that it obscures the political connotations of technological issues. More accurately, we might say that it produces a schizophrenic view of such issues: one that maybe acknowledges the importance of technology in general terms, but which, when it comes down to specifics, sees the politics of technology as being all to do with the choices made at the point of application. This view is reinforced by the 'promiscuous utility' (Winner 1986: 6) of technological artefacts. That they can be used in so many different social contexts seems to sever any link between their design and use.

In contrast, viewing technology as a process – involving not just hardware but also flows of knowledge embodied in or associated with that hardware – makes it possible to see the political factors that link the context of use to that of invention. By determining the flows of knowledge from which technology emerges, and by shaping the context for its use, powerful groups are able to insert their own interests into the technology process. Political interests, in Marcuse's words, are 'not foisted upon technology "subsequently" and from the outside; they enter the very construction of the technical apparatus' (Marcuse 1968: 223).

The influence of powerful groups is one element of the circuit of power relations impinging upon the technology process. But, there are also opportunities for a wide variety of political interests – and not simply the explicit power of the organizational hierarchy – to shape the knowledge and information from which technologies are assembled.

A couple of examples will serve to illustrate the diversity of interests which may figure in this circuit of power relations. First, at the level of organizational contingencies, Pettigrew (1973) describes the influence exercised by a computer manager over a mail-order company's decision to purchase a particular make of computer. This influence is seen as deriving from

1   being able to present oneself as an expert;
2   acting as a 'technical gatekeeper' to control the flow of information between senior management and outside would-be suppliers; and
3   knowing the system, that is, understanding the balance of power within the organization, and how to get things done.

Given his 'boundary-spanning' managerial position, the computer manager was able to exploit these skills to achieve his desired objective.

On a much broader plane, Cockburn (1983) reminds us that 'non decision-making' – the way in which certain groups are largely excluded from the technology process – is itself a political factor. She notes how, in the printing industry until very recently, technology was designed by and for men. This meant that at the material level alone – that is, the physical force needed to operate machines – it largely excluded women from taking on printing jobs.

The relationship between technology and power is not simply one-way, however. Certainly, the technology process tends to reproduce the broad pattern of pre-existing power relations, hence the notion of a circuit of power. Access to the technology process may be limited, as we note in Chapter 7, to the knowledge and information provided by a small range of élite and expert groups, and in that sense technology may become a servant of power. But equally, as we describe in Chapter 3, there are also powerful incentives – especially in a context of innovation and change – to open up the process to a wider range of groups, contexts and knowledges. And as we have already noted, the development of the technology process may open up unexpected contingencies of knowledge, skill and position which can be exploited – as in the Pettigrew study – to advance certain interests and suppress others.

It follows that attempts by powerful groups to unilaterally impose their intentions on the technology process are fraught with tensions and contradictions. Attempts to develop technologies for control may only serve to generate resistance. Or the de-skilling of one group succeeds in re-skilling and empowering another, and so on. In subsequent chapters we will explore some of these contradictions by reviewing, first, the general patterns of technological development in industry, then, the distinctive forms of expertise associated with managerial control of technology, and finally, in Chapter 3, the political role played by users of technology.

Chapter 2

Technology, power and organization

INTRODUCTION

A major influence on the design and use of technology at the workplace is the web of power and dependency relations which links the workplace to powerful groups – technical experts, shareholders, suppliers, trade unions – both inside and outside the organization. This web influences the immediate context for the use of technology, and also helps to shape the interpretation and assimilation of the broader processes of technological change.

There are, for instance, powerful economic pressures to exploit new technological knowledge and artefacts, the consequences of which are discussed in Chapter 8. However, because organizations are not unitary but are made up of different interest groups, such pressures are far from unambiguous in their effect or interpretation.

This is particularly important in the context of management–employee relationships. For example, at the product-market level, management may wish to develop production systems to respond to or promote market growth and differentiation. In ideal terms, therefore, they have an incentive to exploit to the full the knowledge and adaptability of their workforce and to base their design and use of technology on those qualities.

But at workforce level, the economic incentives are not so clear-cut. Management and workers are not simply elements of a functional team co-operating to meet the 'needs' of the production system – though many managers might like to see it in these terms. Their relationship is not only functional but also contractual; to put it bluntly, labour is hired from a labour-market in return for payment. This relationship creates conflicting incentives for each contracting party, where each seeks to define the 'wage-effort bargain' (Baldamus 1961) to their own best advantage. The resulting potential for bargaining and opportunism in the application of labour means that the latter, and the forms of job design and work organization associated with it, is subject to ongoing processes of control as much as to the functional dictates of adaptability and initiative created by the product-market.

Caught between the Scylla and Charybdis of product-market and labour-market pressures, management have to develop strategies to cope with contradictory requirements, namely: 'to limit the discretion which the workers may apply against its interests; and to harness the willing application of that discretion which cannot be eliminated' (Hyman 1987: 40).

Management organize, if not resolve, the problem of conflicting goals by articulating a hierarchy of three broad levels of control, and of management, at which the different goals can be pursued quasi-autonomously (cf. Storey 1985, Buchanan and Boddy 1983, Batstone et al. 1987, McLoughlin and Clark 1988). These are:

1   *Labour control*: the relationship between operational demands and workforce behaviour;
2   *Operational control*: the relationship between strategic demands and the quantity and quality of outputs generated by a particular production system; and
3   *Strategic control*: the relationship between the outputs from an organization's portfolio of production systems and product-market demand.

At each level, a variety of specific 'means' of control (Storey 1985) may be developed, ranging from detailed rules and procedures to much looser, decentralized forms of control. But, in general terms, the nature of such controls, and hence of the design and use of technology, seems to be as much to do with the ensemble of interactions and contradictions *between* the different levels as with the problems *within* a particular level of control (Buchanan and Boddy 1983). A simple example would be the implications which the pursuit of product quality as a strategic goal might have for the conduct of operational and labour control (Batstone et al. 1987). In the guise of programmes of 'Total Quality Management', for instance, such a goal is likely to involve significant changes in shop-floor work practices and job design.

Having said this, in the wake of Braverman (1974), most critical attention has tended to focus on labour control alone. Braverman claims that under capitalism technology is used to reduce the power of the workforce in the production process by centralizing knowledge and control in the hands of management. This view suggests that the general pattern of technology and work organization in industry corresponds to Taylorist or Fordist models – outlined below and also in Chapter 4 – which are seen as all about controlling labour.

The Taylorist/Fordist model

Taylorism and Fordism are said to have de-skilled the previous craft-based forms of work, and cheapened the labour-force, by instigating the following division of labour:

1   Work is fragmented into low-skill, repetitive jobs, thereby expanding the potential labour-market for the employer, and reducing the cost and market power of the labour-force;
2   Workers are subject to close control and monitoring with the aim of maximizing their productivity;
3   There is a separation between the conception and the execution of work; the design and planning of work is carried out exclusively by management.

In this view, Taylorism and Fordism are conflated together. The major difference between them is held to be Ford's substituting an assembly line and technological means of control for Taylor's elaborate hierarchies of direct supervision. Not only did the line reduce 'unproductive' worker movement by bringing the task to the worker, it also

> resolved technologically the essential first control system task; it provided unambiguous direction as to what operation each worker was to perform next, and it established the pace at which the worker was forced to work. . . . Whatever the sequence of tasks chosen, the new setup transformed that order into technological necessity.
>
> (Edwards 1979: 118)

And even now, emphasizing the historical continuity and persistence of these models of control, some of the latest technological and organizational developments seem to be direct descendants of the assembly line. In the retail sector, for instance, the development of the supermarket form of work organization has enabled shop work to be broken down into specialized elements of stock control, pricing and so on, while the customer queue provides a useful pacing effect on the work of check-out operators. Now, the advent of IT in the form of EPOS (Electronic Point of Sale) technology enables even greater management control over the workforce (Child *et al.* 1984). Sainsbury, one of the UK's biggest supermarket chains, have installed laser-scanning tills which monitor the number of items checked out per minute. This has allowed the company to establish a 'comfortable average' of twenty items per minute to be achieved by check-out operators. The company says that it will not discipline workers failing to achieve this norm, but that new operators who did not achieve this level during their induction period would be asked to transfer to other duties (*Financial Times*, 9 May 1989).

Moreover, the linkage between information and communication technologies allows management control over time and behaviour to be extended to an ever-larger range of work-groups. In certain areas of work such as sales, real estate and insurance the workplace can be extended spatially, turning consumer status symbols such as the car phone into a kind of production technology. Wallace, for instance, notes how 'in gridlocked Los Angeles, California, about 200,000 cellular car phones change driving time into productive time' (Wallace 1989: 366).

The irony of the car phone and the supermarket too is that technologies which are seen as representing freedom of choice for the consumer come to embody forms of control and constraint for the worker. This is the essential irony of Fordism itself, where a product technology – the motor car – which greatly enhances the personal mobility of the consumer is created by a process technology which seeks to minimize the mobility of the worker.

Yet the ironies of Fordism also point up some of the features that have made this strategy more than just a means of controlling labour, but also a 'keystone ... of prevailing twentieth century concepts of human management' (Emery 1978: 83). Certainly, Fordism was in part to do with the control of labour; it is surely significant, for instance, that the adoption of the assembly line in Ford's Detroit plant in 1913 coincided with an attempt by the local manufacturers' association to suppress trade unionism in the city (Rae 1965). But, unlike Taylor, Ford's innovations also looked to the wider market-place and to the operational and strategic problems that lie beyond the shop-floor point of production. The power of Fordism lay in its unique, if always contradictory, ability to connect these levels of control through a regime of mass production.

Not only did Fordism provide for a particular form of labour control, but at an operational level the assembly line itself became the key to replacing the complexities of batch production with the simple organizing principle of flow production. Stock and production control were simplified both by the line and by a product range that – famously – extended to one model and one colour. Similarly, the development of specialized machinery dedicated to that one product allowed enormous economies of scale to be achieved.

In turn, such operational features would have been futile had they not been connected to a crude, but effective, marketing strategy. Ford's ability over the first decade of production to reduce the cost of the Model T by two-thirds in real terms (Womack et al. 1990) helped to open up a mass-market for cars. And in that nascent market, features such as standardized parts which made the car easier to build also made it more attractive – since it was thus more readily repairable – to the customer.

As Abernathy (1978) notes, the US motor industry's subsequent development was based on such critical connections between process and product technologies. And as the mass-production regime developed, the objective control it placed in the hands of management allowed the further evolution of the 'product-process life cycle'. A pattern of incremental product innovations could be linked to the advances in process efficiency achieved by mechanization and by the organization of production into specialized engine, assembly and components plant.

Where Fordist strategies have flourished in other industrial sectors, it is often attributable to the effectiveness of this product-process logic. Thus, to return to the retail example, the mushrooming in size and number of supermarkets reflects not only labour control considerations, but also their ability to exploit increasing levels of consumer mobility. In effect, the car frees consumers from dependency on local shops. It then allows them to transport large volumes of goods back from burgeoning out-of-town super-hyper-mega markets. In turn, the latter are able to combine product variety with the kind of low costs achievable through massive economies of scale. Thus Fordism communicates itself not only as a management ideology, as we discuss in Chapter 3, but also through the incestuous interconnection and 'entrenchment' (Colling-ridge 1980) of Fordist technologies in different industrial sectors.

But, while these instances illustrate the power of Fordism in certain contexts, they also help to explain the constraints on its diffu-sion, and the emerging view – characterized by Piore and Sabel (1984) – that Fordism's significance as a 'paradigm' of technological development is in irreversible decline. The effectiveness of Fordism, it seems, is contingent not only on its ability to impose a particular form of labour control, but also on its ability to link that control to the operational and strategic elements of the formula. To the extent that the latter are dependent upon a particular set of assumptions about product-markets and labour-markets – specifically, levels both of mass consumption and of unemployment – any major challenges to those assumptions might challenge the operational and strategic base of the Fordist approach, and thereby threaten radical changes in management practices.

Straws in the wind of such change have even been observed in the motor industry itself. In the UK industry, the Fordist regime has, as Friedman (1977) notes, been periodically undermined by changing product-market and labour-market conditions. In certain periods and for particular groups, Fordist 'direct control' of labour has given way to what Friedman terms a 'responsible autonomy' strategy, involving the creation of a co-operative accommodation between management and workers. This applies with no less force to other industries. Kelly's (1985) study of job redesign initiatives in UK elec-trical engineering companies found a number of firms responding to product-market change by moving away from Fordist methods of job design.

In certain sectors, operational pressures may actually encourage greater autonomy for the workforce. In the branch networks of the UK banking sector, for example, Smith and Wield (1988) argue that the use of new technology for the 'degradation' and routinization of counter-teller jobs (Child 1985) may engender significant oper-ational disadvantages when compared to the traditional paternalistic style of management. The latter has offered strong internal labour-markets and comfortable levels of pay and job security in return for flexible, obedient and trustworthy conduct from bank employees. The attempt by some UK banks to replace this model with Fordist forms of work organization has run into severe problems. It seems that the uncertainty and complexity of banking services make them ill adapted to Fordist or Taylorist methods. Smith and Wield claim that 'specialization tends to reduce staff flexibility. The qualitative and quantitative variations in the "banking day" are inconsistent with Taylorism which works best on continuous inputs and outputs, for example, as occurs on an assembly line' (Smith and Wield 1988: 103). As a result, 'The management, administrative and co-ordination input is increased with specialization despite reductions in branch managers ... there are few if any clerical savings' (Smith and Wield 1988: 103).

FROM FORDISM TO FLEXIBILITY

The dysfunctions of the Fordist model have been aggravated in the last decade by broader social and economic changes. Product-markets have been affected by exogenous political and economic 'shocks' and have become increasingly fragmented and differentiated. The rate of product and process innovation has increased. At the same time, the levels of education and social aspirations of the workforce have led to increasing resistance to Fordist management styles (Piore and Sabel 1984).

In such circumstances, the Fordist approach becomes less tenable. By encouraging long production runs and discouraging radical product change, it privileges process efficiency and product standardization over innovation (Abernathy 1978). In addition, Fordist production systems can no longer be easily protected against product-market change through the 'just-in-case' provision of buffer stocks (Clark and Staunton 1989).

In short, where Fordist strategies seem increasingly rigid, new forms of strategic thinking make flexibility their centrepiece. Product-market change needs to be reflected in a radically new approach to work organization and production technology:

> Flexible specialisation is a strategy of permanent innovation, accommodation to ceaseless change, rather than an effort to control it. This strategy is based on flexible – multi-use – equipment; skilled workers; and the creation, through politics, of an industrial community that restricts the forms of competition to those favouring innovation. For these reasons, the spread of flexible specialisation amounts to a revival of craft forms of production that were emarginated at the first industrial divide.
>
> (Piore and Sabel 1984: 17)

This view highlights differing approaches to the control of labour as a key element in the Fordist–Flexibility divide. These are summarized by Matthews (1989) in Table 2.1.

But while Flexibility strategies are a powerful ideological antidote to Fordism, their adoption at the level of labour control is unlikely to be a direct response to product-market and technological change alone. Indeed, viewing such strategies in terms of different levels of control highlights the extent to which, as our earlier discussion indicated, product-market change may be satisfied by means other than changes in shop-floor work organization. Increasing demand for innovation and product variety may be met, for instance, not by enlarging the product range or encouraging shop-floor flexibility but simply by changing product offerings gradually over time, deleting some lines and adding others (Kelly 1985).

Table 2.1   A comparison of Fordist and Flexible employee relations

Fordist system	Flexible system
1 Management assumptions about workers:	
Worker cannot be trusted and must be controlled by supervisory hierarchy.	Worker desires challenging job and wants to make a creative contribution.
2 Management structures:	
Top–down hierarchy of command with worker simply obeying orders.	Flexible and flat decision structure, emphasizing co-ordination rather than control, and based on project teams.
3 Job design:	
Work is fragmented and de-skilled; jobs narrow; conception divorced from execution.	Work is integrated vertically and horizontally; multiskilled and performed by teamwork.
4 Skill formation:	
Skills are 'bought' from an external labour-market; workers are replaceable and made redundant at economic downturn; little training, no career path; jobs defined by operations and machines.	Skills are nourished by an 'internal labour-market'; job security; training as part of the job; jobs defined by skill level.

Source: Matthews 1989

However, highlighting the possibility of organizational and technological adaptations outside the point of production is to qualify
rather than to reject the Flexibility thesis. As we saw in the banking
example, there remain powerful arguments for the promotion of
Flexible strategies in a number of industrial contexts. However,
achieving the benefits of such strategies depends not only on technological advance but also on the responses of management and
workers, and the relationships between them. Managerial reactions
to flexible thinking and flexible machines are discussed later.
However, at the level of labour control, one of the key propositions
of the Flexibility thesis is that control does not have to be a zero-sum
game, and that both managers and workers can benefit from the
kinds of job design and work organization that in the 1970s might
have been described as 'job enrichment'.

FLEXIBILITY AND THE CONTROL OF LABOUR

On this last point, the evidence suggests that any notional mutuality
of benefits rests ultimately on the perceptions and politics associated
with the bargaining relationship between management and workers.
An example of some of the tensions which that relationship creates
is provided by Wilkinson and Oliver (1990), who describe how even
Ford UK are attempting to adopt more flexible working practices at
shop-floor level. However, certain elements of that flexibility,
notably the minimizing of buffer stocks associated with 'Just-In-
Time' stock control, tend to increase management's dependency on
the workforce. When Ford UK was hit by a strike in 1988, the effect
was felt within days at Ford's other European plant who were
dependent on UK production.

Also, perceptions as well as politics may influence assimilation of
new flexible practices. Flexibility is a managerially determined
response to technological and market conditions. Hence it may
conflict with the workers' perceived interests. For example, 'skills'
have traditionally been understood in the UK as an element of
collective worker control over the work process and not as being in
the gift of management. Thus, if asked, a craft-worker might see
flexibility and autonomy as to do with freedom from management
control rather than a variety of job tasks. Such autonomy might
encompass the ability to resist the employer's demands for mobility,
for instance, or even the ability to prevent other work-groups taking
on a wider variety of tasks which conflict with craft demarcations. In

contrast, the kind of flexible working that the employer pursues may seem to be directly contradictory to the worker's ability to exercise this kind of autonomy, and may often involve overriding previously established craft-based controls (Scarbrough 1984).

The perceived conflict between management's concept of flexible skills and established craft controls is expressed by one of the craftsmen at Ford UK, who gave this account of what Flexibility may mean to the individual worker:

> I was taken on as a millwright. I now have to double up as a rigger, welder or plumber. The training is a joke – we had three days learning welding skills.... Now they want to shove us on the lines when they want. It's not on.
>
> (Quoted in *The Independent*, 9 February 1988)

And although the flexible strategy may emphasize the importance of up-grading knowledge, 'skills' and autonomy these may be seen by the worker not as enhancing flexibility but as increasing their workload and responsibilities. Indeed, production management may seize on flexibility initiatives to do exactly that, by opportunistically extending the range of tasks demanded of production workers. In the early 1980s, when British Leyland imposed a new set of flexible working practices on their shop-floor workers, one of the foremen at the company's Longbridge plant enthused: 'We can move workers about anywhere now. We have them sorting out scrap and cleaning up their section' (Scarbrough 1982: 81).

If truly flexible forms of working are to be developed, given the kind of constraints noted above, an important element is likely to be the achievement of a qualitatively different form of employment relationship between employers, managers and workers. This is likely to be one in which the worker is socially and not merely functionally integrated into the organization, such that the subjective or collective perception of flexibility is at one with that of management.

So far, the achievement of such a unitary relationship in the UK at least is highly questionable (Pollert 1988). And even where it is achieved, it seems likely, as far as the workforce is concerned, to be double-edged. On the one hand, multiskilled 'core workers' may be more fully integrated into the organization (awarded staff status, and so on), but the same functional and economic logic that lies behind the social integration of 'core workers' may simultaneously require the, even greater, social exclusion of 'periphery' workers who possess no company-specific skills. The latter groups may continue to experience Fordist or Taylorist forms of management control emphasizing temporary contracts at the lowest possible wages (Atkinson 1984). To the extent that such strategies are pursued, particular groups – women, racial minorities – who are disproportionately represented in the 'peripheral labour force' may well find that flexible technology in one area means increasingly rigid forms of control in another.

On the management side, too, the changes attendant on Flexibility are no less far-reaching or politically sensitive. The managerial perceptions and understandings generated by existing structures of control seem likely to conflict badly with the quest for 'permanent innovation' or quasi-autonomous groups at shop-floor level. Such a conflict is already apparent in the ranks of middle management who actually implement programmes of job redesign and work organization, as the empirical experience of quality circles and other teamworking initiatives demonstrates:

> There is a considerable body of evidence to suggest that middle managers and supervisors are the biggest stumbling block to the introduction of quality circles, and that they are largely responsible for many of the quality circle failures.
>
> (Collard and Dale 1989: 375)

More generally, Flexibility strategies call into question the control of knowledge exercised by the many and various specialist functions within management. Perhaps the most powerful legacy of Taylorism and Fordism, in fact, is not the control of shop-floor workers *per se*, but the equating of power and control with specialist and objective forms of knowledge. Piore and Sabel's notion of Fordism as a 'paradigm' gives a sense of this model being embedded in the thinking of management. However, it is also embodied and institutionalized in formations of experts and specialists. As Kuhn's (1962) work on paradigms in science makes clear, the latter are likely to need more powerful incentives than just 'objective' evidence of Flexibility's advantages to change their way of thinking. Thus the concentration of knowledge in the hands of expert groups has important repercussions on the conduct of management in general, and on management–worker relations in particular.

THE POLITICS OF EXPERTISE

Although Fordism encouraged the growth of expert and professional groups within management, the problems of controlling knowledge-workers proved not dissimilar in certain respects to those of controlling skilled craft workers. Although the knowledge-workers constituted a new form of management structure, experience was to show that such a structure was neither monolithic nor simply a tool of capital. Intramanagement politics encouraged competition between different specialisms to secure control of the most important decision-making functions.

In this competitive context, the fate of the engineering profession is particularly illuminating of the kind of strategic knowledges that have come to dominate the technology process. Although both Taylor and Ford had determined that engineering knowledge was a critical component of technology and organization, engineers as an occupation were unable to fulfil these expectations. In part, this was precisely because Taylor and Ford had succeeded only too well in codifying and organizing production engineering knowledge with the result that, as Armstrong notes, the 'techniques of scientific management proved too lucid and could too easily be detached from the ambitions of the engineers' (Armstrong 1985: 132). At the same time, the emergence of higher levels of control in multi-divisional organizations encouraged the emergence of finance and accounting specialists, who could claim that 'decisions of allocation between dissimilar operations could only be made on a common, abstract – and therefore financial – basis' (Armstrong 1985: 136).

The dominance of accountants over engineers, which is still a feature of UK and USA industry, highlights the politics of expertise in the technology process. The determination of the forms of know-ledge that shape technology is arrived at not by objective 'scientific' means, but through competition between the claims to knowledge made by different groups. Such claims are promoted by the emer-gence of occupational and organizational specialisms. The latter devote themselves, in two important respects, to the control of knowledge. First, at a functional level, they provide a means of codifying and transmitting knowledge through formalization and the promotion of standards. Second, in more political terms, they aim to advance the interest of their members by promoting exclusive control over particular domains of knowledge.

To an extent, it is impossible to separate the political from the functional in the claims made by such specialisms. The means of generating, validating and transmitting knowledge – examination structures, apprenticeships (Keep 1989) – are useful to employers inasmuch as they serve to generalize and externalize the costs of expertise. Yet, they serve also as the means of demarcating and restricting access to the knowledge-base of the specialism. Similarly, the outward mystery and indeterminacy of the occupation's claims to knowledge may be both a reflection of the inherent uncertainty and complexity of knowledge, and a means of advancing its political interests.

At the same time, however, the greater the power and status of a particular occupation the greater the incentive for organizations (or specialisms within them) to seek to abstract that occupation's knowledge from its social and political roots. This may occur – as with the engineers – through the codification of occupational knowledge allowing other groups to 'cannibalize' it for their own purposes. Or knowledge may be embodied in technological artefacts and commodified – turned into a product – making it more readily accessible. Developments in computer software and expert systems, discussed in Chapter 7, which facilitate this abstraction of knowledge, serve to both empower computing experts and simultaneously undermine the expertise of other groups. The air traffic control example, outlined below, illustrates the political consequence of such an abstraction in a context where the expert group's own 'professionalism' effectively connived with management in rationalizing it out of existence.

Case example: Technology and air traffic control expertise

Having long resisted calls from the Professional Air Traffic Controllers' Organization (PATCO) for a computerized air traffic control system to replace outmoded equipment, management at the US Federal Aviation Authority (FAA) finally installed a new 'flow control' system. This used computers to spread out the sharp concentration of arrivals and departures they had previously tolerated in daily operations.

Subsequently, in 1981, PATCO's 11,500 members took strike action over pay and working conditions. However, during the dispute FAA management were able to keep the system going without the aid of the striking controllers thanks to the new technology. The sophisticated computer-driven 'flow control' system effectively flattened the peaks and troughs of air traffic activity and enabled the FAA to handle 83 per cent of the system's former traffic load with less than 50 per cent of its pre-strike workforce.

Hence the technology contributed to a complete undermining of PATCO's expertise and bargaining position. The dispute reached its climax in October 1981 when FAA management decertified the union, with the full support of the newly elected US President, Ronald Reagan.

Source: Shostak (1987)

A related means of dealing with expertise is to subject an occupation's claims to knowledge to the kind of commercial 'Occam's razor' provided by market forces. As we note in Chapter 7, this can range from the 'buying-in' of certain forms of specialist expertise to the development of market controls over in-house groups.

The alternative to such strategies – where an occupation's knowledge-base is seen as too central to be 'marketized', for instance – is to reduce any potential organization–occupation conflict of interests through the social integration of knowledge-workers into the organization. This involves recognizing and rewarding them as 'core workers' in the language of the Flexibility thesis.

However, there remain a number of groups whose expertise is not readily controlled through either markets or hierarchies, and who consequently exert the most explicitly occupational influence on the technology process. Groups which most clearly come under this heading are those whose knowledge is valued either directly or very nearly for its own sake. This may encompass academic and research scientists, and even certain computing specialisms where technological knowledge is changing too rapidly to be properly organized and codified.

Reflecting the indeterminate nature of their professional knowledge, where means and ends cannot be tidily divided, it becomes even more difficult to divorce the productive knowledge of such groups from their political interests. They may become the arbiters of their own work in a way which ensures considerable professional autonomy in whatever organizational context they operate.

One of the political consequences may be that the technology process of the organization or industry becomes subject to the claims of occupational expertise, and not vice versa. This produces what we can term a 'power loop' – see Figure 2.1 below – in which a powerful expertise is able to reproduce itself by maintaining complete control of the technology process.

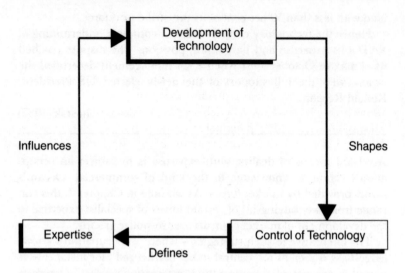

Figure 2.1 The technology power loop

Two of the most striking examples of such a power loop in operation come from the medical profession in the UK and the USA. In one instance, described by Child *et al.* (1984), a system of computerized medical diagnosis was introduced into a UK hospital. This system, which diagnosed hypertension, could have been operated effectively by nursing staff, or even by patients themselves given a few simple modifications. That it was not used in this way was attributed to the professional interests of the consultant in charge. He 'maintained total control over the process of applying the new technology' (Child *et al.* 1984: 167). Furthermore, the argument for the medical profession's exclusive control of the technology rested on a classically self-validating claim: that expert medical judgement was still needed to gauge the truthfulness of patients' responses.

The political corollary to controlling a particular technology process, of course, is suppressing it. This power too have been exercised effectively by the medical profession as the PROMIS case example below demonstrates.

Case example: PROMIS and the politics of medical expertise

PROMIS (Problem Oriented Medical Information System) was

developed in the USA during the 1970s in order to apply a computing solution to some of the problems inherent in medical care within large hospitals: notably, poor organization of medical records, and over-reliance on the memory of doctors. PROMIS incorporated computerized relational databases for medical records and automated sophisticated administrative functions. Access to these functions was via a touch-screen terminal. Yet, despite its information-processing potential, the adoption of the system has been very slow. Reasons for this emerged from the findings of experimental implementations within a number of hospitals in the latter half of the 1970s. These indicated the important role played by the politics of expertise.

It was found that nurses welcomed the openness of the system as it expanded their professional discretion and ability to intervene without approval from doctors. Pharmacists and radiologists welcomed the system for similar reasons. It was the doctors – the most powerful expert group within the pre-existing balance of power in the hospital – who proved most resistant to the technology. PROMIS impinged on their sovereignty by codifying elements of their knowledge-base and ceding some of their expertise to other expert groups.

This potential technological threat was seen to be blunted, however, and the doctors' autonomy preserved, by continual advances in medical knowledge. These maintained the indeterminacy of the doctors' knowledge-base and effectively outstripped the ability of systems such as PROMIS to capture and codify it. The implications of continued dependency on this human form of medical expertise, together with the key organizational positions held by senior doctors, not surprisingly served to greatly retard the uptake of the PROMIS system.

Source: Willcocks and Mason (1987: 15–16)

SUMMARY

We have noted the powerful hold that Fordism has exerted upon technological developments in key sectors such as the car industry. We have also noted the political and strategic issues still to be resolved before the Flexibility model can be said to have attained such paradigmatic effect. In particular, the tendency for expert groups to reproduce their power and influence within the technology process, or, indeed, to expand it at the expense of other groups, does not bode well for the diffusion of knowledge and control implied by more flexible, non-bureaucratic forms of working. More specifically, the concentration of knowledge and power in the hands of a small set of expert groups conflicts with the political and functional pressures towards greater user participation in the technology process: a conflict that we explore in details in the next chapter.

Chapter 3

Technology, power and the user

INTRODUCTION

At the organizational level, the domination of the technology process by managerial and specialist forms of expertise seems to leave little room for user participation. Nor do the traditional arrangements of collective bargaining, at least in the UK, offer great scope for workers to exercise influence over the design of technology (Martin 1988a). This applies not only at company but also at plant level: Daniel (1987) reports from his survey of UK workplaces that shop stewards were only formally consulted on technological change in a minority of cases – 39 per cent in the case of manufacturing firms, 20 per cent for services. In short, as Martin puts it: 'Trade unions have exerted little influence upon the pattern of technological change' (Martin 1988a: 123).

But if such findings demonstrate – as writers such as Braverman (1974) would assert – that the technology process is predicated upon the unequal power relation between capital and labour, in the workplace that same process also seems to reveal some curious tensions and contradictions which make it more than a trivial or routine exercise. The absence of technological design from the collective bargaining agenda, for instance, needs to be balanced against the amount of attention that managers give to the design of the use process for technology and the work practices that are embedded in it (Martin 1988b). This is not simply a matter of gaining worker co-operation in the use of technology, though that is an important issue which will be discussed later (see pp. 64–8). Rather, there is evidence that technological development in general is actually facilitated by some kind of user participation (Willcocks and Mason 1987, Majchrzak 1988, Corbett 1989).

Type of participation

Phase of technology process		FORMAL	INFORMAL
	DESIGN	1 Trade union consultation Prototyping	2 User redesign
	IMPLEMENTATION	3 New technology agreements Collective bargaining	4 Skills bargaining Negotiation User co-operation
	USE	5 Job design Quality circles	6 Informal job redesign and work practices

Figure 3.1 User participation in the technology process

Whatever the broader circuits of capital–labour inequality or pre-existing bargaining positions, as the technology process unfolds it enjoins managers and workers alike to seek influence by using every available opportunity to insert or withhold particular pieces of knowledge or information. As McLoughlin and Clark put it, 'Worker influence, and that of managers too, often depends less on traditional bargaining resources than on the differential degree of information and knowledge of the new technologies of the different actors involved' (McLoughlin and Clark 1988: 94).

The incentive to exploit knowledge and expertise as a means of shaping the technology process – as we saw in the example of the medical profession – is that failure to do so may lead to their irretrievable loss. For, as Martin notes, 'the major resource redistributed by technological change is knowledge; groups with knowledge of the old system may lose control of knowledge under the new system' (Martin 1988a: 119).

Such factors suggest that the importance of user participation in – or exclusion from – the technology process depends, at least in part, on the contingencies and conflicts that arise at each phase of technological development. The remainder of this chapter is structured, as Figure 3.1 indicates, around the possible contexts for participation, and will explore some of the political implications created by each.

THE DESIGN PROCESS

'Formal user participation in technological design' (box 1 in Figure 3.1) is limited since management will seek wherever possible to incorporate their objectives in the design of organizational technologies such as factory layout and machinery. This not only reflects an attempt to make the production process more objective and transparent, thereby enhancing management control and facilitating innovation, it also corresponds to the underlying rationality of managerial work and philosophy and facilitates the implementation of managerial goals.

The early 1980s saw a significant flurry of activity on the part of UK trade unions in relation to new technology with a number of unions issuing policy statements and the Trade Union Congress (TUC) producing an outline set of guidelines for trade union responses to new technology. In truth the latter contained sentiments that were not entirely different from the kind of policy statement that had been issued in the 1950s in relation to the then current concerns about 'automation': i.e. the demands for consultation over technological change, the notion that the workforce should receive a share of the benefits of technological change, and that change should involve as little unemployment as possible. In effect, the unions' position involves a focus on the extrinsic aspects of technological change – jobs, pay, working conditions – together with a determination to avoid the charge of Luddism.

The significance of the UK experience at formal levels can be gauged by a comparison with research carried out by the European Foundation in Dublin (Cressey 1985). This reveals that, within European business, formal user participation in design is generally restricted to management briefing sessions to trade unions once the design phase has been completed. The participation process was seen as taking place within very narrow technical boundaries.

However, the European Foundation research did uncover ten cases of formal consultation with trade unions over the *selection* of technology. Of these cases, eight involved 'white collar' unions. This Cressey (1987) takes as indicating three things: (a) the lack of an existing collective bargaining/union tradition of regulation, (b) the possibility that the nature of the innovation may be amenable to involvement, and (c) management dependency on the skills and initiatives of this group of workers. Given our earlier discussions of the politics of expertise, it is likely that management dependency on user expertise provides the most compelling reason for granting users a degree of influence over design.

While clearly something of a rarity, the potential for formal trade union participation and/or the participation of users in the actual design of technology is illustrated by examples such as the, perhaps unfortunately named, 'UTOPIA' project in Sweden. This was an ambitious attempt on the part of the Swedish Centre for Working Life, in conjunction with the Nordic Graphic Workers' Union, to achieve two objectives: (a) to evaluate and develop the conditions in which formal participation in the design and development of technology could become a viable trade union strategy, and (b) to uncover realizable technological and organizational alternatives to 'Fordism' and the perceived de-skilling of workers associated with new technology.

The UTOPIA project (UTOPIA is a Swedish language acronym for Training, Technology and Products from the Quality of Work Perspective) grew out of a realization that formal trade union participation in design faced a number of problems. First,

> the degrees of freedom available to design the content and organisation of work which utilises existing technology is often considerably less than that required to meet trade union demands. Existing production technology more and more often constitutes an insurmountable barrier preventing the realisation of trade union demands for the quality of work and a meaningful job.
>
> (Ehn *et al.* 1981: 7)

A second problem was seen as relating to the increasing preference on the part of management for the purchase of 'turn-key' systems designed by experts outside the organization, rather than for 'in-house' system development. The resultant distance (both geographically and attitudinally) between designers and users makes user participation highly problematic. Indeed, Pacey (1983) argues that a cognitive disjuncture between 'experts' and 'users' is a trademark of technical expertise and is reflected in the assumptions made by the former during the design process. These include:

1 assumptions based on academic specialisms and on the need to maintain boundaries between professional experts;
2 a tendency to overlook opportunities for improvements in the use process and go for technical fixes instead;
3 failure to recognize the invisible organizational aspects of technology invariably developed by users which often contribute to a more effective operation of technology;
4 failure to recognize the political nature of technology in organizations – the conflicts of values, interests and social goals that specific technological projects may entail.

The UTOPIA project team (comprising social scientists, computer scientists and trade union representatives) successfully developed a prototype text and imaging processing technology for use in the typographic industry which enabled traditional typesetting craft skills to be utilized and developed through the use of computers.

Three important lessons may be learned from the UTOPIA project. First, it is possible, but highly unusual, for trade unions to engage in the design of new technology. The potential benefits of such participation are revealed by the experimental development of a design process known as 'design by doing' which enabled 'computer-illiterate' workers to make a more meaningful contribution to design decision-making than is possible in conventional systems analysis and design (Ehn 1988). In this context, the contrast between the Nordic Graphic Workers' approach and the stance adopted by the UK Fleet Street print unions with regard to the introduction of computer-based typesetting technology could hardly be more pronounced. But second, it is important to note that the UTOPIA project was an experiment conducted outside the business organization environment and has not led to a widespread shift in trade union strategy in Scandinavia.

Last, despite the pioneering work within the project, a key constraint on formal trade union participation in design became apparent. Workers lacked the necessary expertise to participate fully in the process through which their system specification was translated into a technical artefact by engineering designers and technical experts. The actual manufacture of the UTOPIA system was subcontracted to an engineering company and the resultant system differed in a number of important respects from the workers' original system specification. This difference stemmed from 'technical difficulties' experienced by the subcontractors during the design process. They argued that certain elements of the original system specification were either unrealizable or inefficient in a technical sense. The lack of technical 'know-how' and expertise on the part of the workers made further negotiation extremely difficult and the system was finally built to a specification modified by the engineers.

These lessons highlight the importance of relative levels of expertise in determining the effectiveness of user involvement in design. They are reinforced by the experience of a similar project at British Leyland in the UK, which is outlined in the case example below. One of the implications of both the UTOPIA and the British Leyland projects seems to be that user participation can be made more effective only when the technical design process is itself re-designed (see Ehn 1988 and Corbett et al. 1991 for a more detailed consideration of this issue).

An important contribution to this debate is provided by socio-technical systems theory, which has formed the basis for a number of 'alternative' design approaches, most notably Mumford and Weir's (1979) ETHICS method (Effective Technical and Human Implementation of Computer Systems). However, the evidence shows that socio-technical systems designs are predominantly led by specialist expertise and, as the 'I' (for implementation) in the ETHICS acronym indicates, they tend to concentrate more on fitting jobs to the technology than fitting technology to the job (Taylor 1979, Beirne and Ramsey 1988). Conversely, Checkland's (1981) 'soft' systems approach to the design of technological systems is more firmly rooted in the analysis of organizational practice and says surprisingly little about machines and other physical technologies.

The relative strength of Checkland's approach – vis-à-vis conventional systems approaches – lies in its stress on the qualitative and informal aspects of organizations and, more importantly, in its recognition that conflict may exist. However, the 'soft' systems approach addresses conflict at the level of perceptions and ideas and ignores conflict arising from the politics of expertise and from material and economic interests (Burrell 1983). So, while perhaps not technically driven, the 'soft' systems approach can end up serving the interests of the most powerful organizational groups, thereby perpetuating existing power structures.

In short, at the formal level, the planning and design of new technology remains a predominantly management-driven exercise, with user participation an area of 'non-decision making' (Bachrach and Baratz 1962). Important decisions regarding investment and systems design lie outside normal bargaining processes and rarely, if ever, appear on the formal political agenda. User involvement in technological design generally extends, as the European Foundation study found, to, at best, notification – that is, information distribution and briefing sessions – rather than participation.

Case example: British Leyland and formal user participation in technological design

A rare instance of formal worker participation in technological design in a UK company comes from the workings of the 'AD088' joint management–union sub-committee at British Leyland's Longbridge plant in the late 1970s. This committee, which was composed of shop stewards and managers from the plant, operated under the aegis of a company-wide participation system which had been established at government prompting (British Leyland being then a nationalized company) in 1976. Over a period of four years, it had an opportunity to examine and comment on the new bodyshell assembly technologies that were being developed to manufacture the Mini Metro car (code-named AD088).

A study of the committee's operations reveals two major constraints on the trade unionists' involvement: expertise and efficiency.

Expertise

Trade unionists on the AD088 committee had a rare opportunity to scrutinize design proposals at an early stage. However, to a large extent, the problem for the unionists lay not in the timing but in the 'technical' nature of such proposals. With only an abstract outline of the operations of a particular machine or system, the trade unionists were unable to identify the implications of the new technologies for the work practices of shop-floor workers. Nor did they have the expertise or information to be able to challenge the kinds of design criteria that were being embodied in the new systems. The unionists were therefore placed in the passive position of reacting to management proposals.

Efficiency

An equally important constraint on union involvement was their acceptance of efficiency as a primary design criterion. As management not only controlled the design process, but were also responsible for defining the relative efficiencies of the various technological options, this tended to constrain the trade unions' input. Even the unions' major aim of enriching jobs on the new production lines was secondary to such considerations within the overall technological design. As one of the AD088 committee stewards explained: 'We thought that there should be as much opportunity as possible for operator initiative in the production process, and this would have to *fit in* with the choice of technology' (our emphasis) (Heller and Hitchon 1979: 15).

Despite their inability to challenge the key parameters of the design process, the trade unions were happy with their involvement in the AD088 committee. In the words of the senior steward, it was 'the most effective participation exercise that we took part in ... less abstract than other exercises'.

An example which explains union satisfaction with this particular exercise, but also the terms under which that satisfaction was obtained, is provided by the extensive provision of so-called 'buffer stores' in the design of the new plant. These were, in effect, mechanical magazines into which the workers would feed component sub-assemblies which would then be automatically fed from the magazine and welded by multi-welder machines into an overall bodyshell. By this arrangement, the workers feeding the various welding machines in the new factory were freed from strict pacing by the multi-welder machines. The senior steward argued that this feature embodied the unions' aim that the 'individual should not be abused by the technology'. The steward felt that the half-an-hour 'buffer' for each sub-process meant that 'the bloke has half-an-hour's control over the machinery, and in fact he is controlling the process rather than the process controlling him'.

But, whatever the benefits to the workers themselves – and at least one study has found that there is a direct positive correlation between the size of buffer stores and job satisfaction (Klein 1978) – the pros and cons of this arrangement were viewed in a somewhat more instrumental light by management themselves. Far from seeing the buffer stores as representing the workers' 'control over the machinery', the industrial engineers responsible for machine layouts saw this feature as a means of insulating their highly automated and integrated system against two possible threats to its 'technical integrity':

1   a machine failure in one sub-process area having knock-on effects for the whole system, and
2   the 'uncontrolled interaction' that might result from mixing man-controlled operations (frequent, short duration 'interrupts') with machine-controlled operations (infrequent but long duration interrupts).

In effect, British Leyland's industrial engineers saw the buffer stores as a means of securing the objective integrity of the system against the subjective uncertainty of the workforce. Moreover, by allowing mechanization to be extended into sub-assembly areas, the buffer stores also helped to break down the usual sub-assembly work groups into individual and isolated machine-loaders. As one of the industrial engineers explained it: 'technology breaks down those groups and isolates the workers and turns the operators into automata'. One industrial engineer estimated that the resulting elimination of 'social constraints' on operations would increase machinery efficiency levels by 12 per cent.

'Informal user participation in technological design' (box 2, Figure 3.1) is certainly less well documented than its formal counterpart. Such *ad hoc* user participation in design is more correctly termed 'redesign' as it stems from modifications undertaken by users once technology has been in use for a period of time. For example, it is not unheard of for skilled, numerically controlled (NC), machine tool operators to design and build their own NC tape program generator/editor so that they can rewrite NC programs prepared by office programming staff. This illicit redesign practice, if and when discovered, may even gain tacit management approval if the 'illegal' tapes prove to be instrumental in reducing scrap levels or increasing output or quality.

However, such radical redesigns are rare. Typically, *ad hoc* user participation is restricted to two areas. First, the incremental development, arising from use, of minor technical modifications which result in the smoother or more efficient operation of equipment. These modifications are sometimes simply a response to excessive production demands, 'unrealistic' production targets, or the need to maximize earnings based on piecework. Safety guards which slow down the loading and unloading of tools or products from a machine, question-driven CNC controllers which force operators to respond in a fixed sequence and at a frustratingly slow pace, and quality-testing software which reviews all machine operations carried out on a piece of work can all be overridden by a user with a degree of technical know-how.

The second area of 'adhocracy' is the occasional visit to the shop-floor of in-house technical design experts who, like the earlier generation of time and motion personnel, are keen to observe the production process with a view to increasing output and/or quality. A kind of informal 'knowledge engineering' can occur when a software design specialist observes and interviews a number of unsuspecting skilled workers about their work with a view to codifying and ultimately automating particular aspects of the labour process, for example, error detection and retrieval. Although such a process is often desirable – from a managerial perspective – at the formal level, the threat to the skilled worker's knowledge-base, and hence relative power position, is such that resistance and non-co-operation often ensue. Indeed, the informal use of 'knowledge engineering' offers a good example of expert groups' efforts to extend their own knowledge-base.

THE IMPLEMENTATION PROCESS

The design and development of technology imply the ability to bring it into successful operation. However, as Willcocks and Mason note of IT implementation:

> while the ability of the systems analyst may rarely be questioned in technical matters like analysis and design of hardware and software selection, such expertise and associated technical qualifications may form an insufficient power base for fulfilling the responsibility for seeing the project implemented.
>
> (Willcocks and Mason 1987: 173)

Even in minor applications, existing power structures will be disturbed by the introduction of new technology and, as a consequence, the implementation process will reproduce (and amplify) existing strains and conflicts within the organization. Hence, although management will often attempt to incorporate their production goals into the design of hardware and software, when such designs encounter the local and social realities of the use process, management objectives – and sometimes even the design configurations of the hardware itself – may become fragmented. This may lead to a further series of adjustments and 'sub-strategies' (Clark *et al.* 1988) as front-line management try to rescue the original goals of the technological change.

'Formal user participation in the implementation of technology' (box 3, Figure 3.1) is becoming increasingly widespread. The European Foundation research shows that the implementation phase is characterized by the high use of collective bargaining and formal trade union consultation. Although the use of New Technology Agreements (NTAs) is often a key part of union strategies, research evidence suggests that they are limited in both number and impact (Ruskin College 1985). One reason for this limited impact is that NTAs tend to suffer from their own comprehensiveness and detailed specificity; substantial qualification is needed to adapt such agreements to local organizational and sectoral conditions (Cressey 1987). Even in (West) Germany, where a comprehensive set of national and interorganizational NTAs exist, local collective bargaining is often preferred by all bargaining parties.

Thus, despite attempts in the mid-1980s to deal with technological change through innovations in collective bargaining, evidence tends to support the conclusions of Manwaring's (1981) study that, with the exception of health and safety issues, many of which are bound by various Acts of Parliament, 'unions have been largely unsuccessful in securing a share of the benefits from new technology; the rhetoric of model agreements has not, in general, been translated into negotiated concessions in clauses of actual "new technology agreements"' (Manwaring 1981: 59). But management's customary dominance of the implementation process is not only a result of the formal inadequacies of NTAs. It must reflect, too, underlying constraints on the scope of collective bargaining in general.

For example, in recessionary periods employers may plausibly claim to be unable to deliver on union demands. The so-called trade union 'new realism' of the early and mid-1980s was based in part on a perception that employers are subject to the vagaries of government policy and the peaks and troughs of economic cycles. In the early 1980s, manufacturing industries in both the UK and the USA underwent a period of recession during which large-scale redundancies, attacks on custom and practice and work rules and, especially in the USA, actual wage reductions occurred frequently. And even in the state sector, threats of closure (as at British Leyland), privatization and/or the broader effects of technological change itself – for example, the threat to the Post Office's letters business posed by FAX machines – served to moderate trade union demands and encourage an acquiescent or even enthusiastic attitude to new technology.

A second constraint on collective bargaining relates to competition between trade unions, a classic example of which is outlined in the case example of CNC machine-tools outlined below. In this context, certain unions – represented in the UK, for instance, by the electricians' union, the EETPU – may seek to out-bid others in order to gain employer recognition and hence members. This may include a willingness to pursue 'single union' or even 'no strike' deals with employers in an attempt to guarantee job security and favourable pay and working conditions for union members. Such strategies inevitably undermine the ability of other unions to pose more radical demands on the implementation of new technology.

The 1981 PATCO union dispute with the US Federal Aviation Authority (FAA), which was discussed in Chapter 2, offers another pertinent example. As the dispute took its course, the Airline Pilots' unions sided with the FAA, while the Machinists' Union indicated it would strike to back PATCO only if other relevant unions took similar action. This lack of support from two unions which, alone or together, could have won the strike for PATCO in 24 hours, shut off this route and contributed to the eventual decertification of PATCO by the FAA.

Case example: Computer numerical control (CNC) and inter-union conflict

A classic case of inter-union competition developed in the UK following the development of CNC machine-tool technology. The widespread application of NC (numerical control) technology into industry in the 1950s and 1960s had enabled management to separate planning from doing. These earlier NC machines were controlled by a paper tape containing a series of recorded instructions which were carried out automatically by the machine control system. The preparation of the paper tape was carried out by means of a special stand-alone piece of equipment and necessitated the skilled use of a specialist programming language (APT). These programming language skills were very different from those associated with the operation of manual and semi-automatic machine-tool technology and, as a consequence, control over NC programming was placed in the hands of technical white-collar office staff, many of whom were members of TASS – the Technical and Supervisory Staff union.

But, with the advent of CNC, and with 'user-friendly' programming systems incorporated into the machine tool itself, a series of demarcation disputes ensued throughout the metal-working industry between TASS and the AEU (the Amalgamated Engineering Union) over whose members should carry out CNC programming given the demise of APT and the paper tape. This struggle over 'ownership' greatly weakened the respective unions' bargaining power over new technology as management were able to play one union off against the other.

A third constraint on collective bargaining arises from employers' attempts to avoid the recognition of union bargaining rights by pre-empting the spread of union membership among employees. Indeed, this avoidance of union recognition seems to be a particular feature of many 'high technology' companies, with IBM an egregious example (Dickson *et al.* 1988). The effect of anti-union policies among high-tech companies is reinforced by more general changes in employment, which tend to favour industries and areas where union organization is weak. These structural changes include the shift towards service sector employment in virtually all western economies, the movement of working population from the 'rust-belt' to the 'sun-belt' in the USA, and the rise of white-collar, non-manual occupations.

As in the case of design, formal participation in the imple-mentation of new technology is also constrained by the 'closed atti-tudes' of both management and trade unions. Given the uncertainties generated by technological change, management's 'default option' often involves regaining control by delegating responsibility to technical specialists who would then impose the necessary solution. Cressey observes that

> overall, the tendency to closed attitudes resulted from the appli-cation of a 'mechanistic' response to the problems of change and the management's fear of a lack of control. This then tended to

introduce rigidities of approach to the organization of innovation, possibly at a time when change dictated the very opposite.

(Cressey 1987: 15–16)

Trade unions often exhibit equally closed attitudes towards the possibility of formal participation. Historically, unions have often taken a reactive stance to change, such that when they do take part in issues circumscribed by managerial influence role confusion often ensues. For example, trade unionists are understandably reluctant to participate in the kind of restructuring often associated with technological change which entails detrimental effects on jobs or work reorganization for some of the union's membership. In general, then, as Kelly (1985) and Beirne and Ramsay (1988) indicate, the potential benefits of such participation for trade unions are seen as being outweighed by the potential for management exploitation and advantage. A particular danger is the possibility that such exercises may expose workers to management's expert power and thereby give them, at best, a false sense of involvement.

'Informal user participation in the implementation of technology' (box 4, Figure 3.1) is crucial for the successful installation of technology in organizations. Although the pursuit of managerial control and the reduction of the subjective element may be important elements in the design of new technology, there is a great deal of evidence to indicate that in use all technologies involve some degree of worker co-operation. Even highly controlling technologies such as the assembly line depend ultimately on the workers doing more than simply carrying out those elements in the task that are pre-ordained or determined by the supervisor. For example, so-called 'unskilled' workers use their initiative and 'tacit skills' to cope with unpredictable variations arising during production (Kusterer 1978). And other technologies may need to enlist more active forms of co-operation involving the application of high levels of skill or a positive commitment to organizational or functional goals. This need is likely to be accentuated by the persistence of high degrees of uncertainty and complexity even in highly automated plant.

Batstone *et al.*, for example, describe how the management of an engineering company were heavily dependent upon the skills of their manual labour-force, and consequently decided to opt for shop-floor programming of CNC machines rather than having it carried out by planners in a central office. 'Any challenge to the autonomy of the craftsmen it employed could lead to serious consequences, not only in terms of industrial disputes but, much more important, in more subtle forms of withdrawal of goodwill' (Batstone *et al.* 1987: 215).

Because of such dependence on user 'goodwill' and co-operation, managers in some companies will engage in subtle and informal forms of negotiation with workers to smooth the way for technological change, knowing full well that failure to do so makes the new technology highly vulnerable to what Wilkinson (1983) terms the workers' 'subterranean' means of influence. Thus, as Batstone *et al.* found from their overall analysis of case study material drawn from the brewing, small-batch engineering, chemicals and finance sectors, 'industrial relations considerations significantly shaped management's approach. Changes need either to be acceptable to employees or to be compensated for in some way if confrontation or a lack of cooperation is to be avoided' (Batstone *et al.* 1987: 216).

But, while industrial relations factors may make some managers keen to secure the co-operation of their employees, the same factors may have quite a different effect on other managerial groups. Indeed, in some companies management may themselves resort to subterranean means to construct the appropriate political conditions for technological change. Secrecy and the careful timing of change are obvious ways of turning management's informational advantage over the workforce into a political one. This kind of political factor has led at least one producer of industrial robots to disguise its technicians as 'Heating and Ventilation Engineers' when evaluating potential shop-floor installations (Scarbrough 1981).

A similar exploitation of informational advantages may involve the implementation of technology under cover of an artificially created crisis. The 'crisis' is used to legitimate the introduction of new technology and/or new working practices as a part, or all, of the 'solution' (Pettigrew 1985, Harvey-Jones 1988).

Finally, Wilkinson (1983) usefully summarizes the role of shop-floor workers in the technology implementation process by arguing that, New Technology Agreements notwithstanding, bargaining over skills and the organization of work generally remains at an unofficial, often covert level. This is not only because technological change generates a degree of uncertainty which cannot be accommodated within existing power structures. It also has to do with the shrouding of such change in an ideology of 'efficiency'. This tends to set a 'frame' around decision-making such that the often political grounds on which choices are made are effectively concealed during negotiations. As we discuss further in Chapters 4 and 5, the ideology of efficiency plays an important role in shaping the responses of managers and workers alike to technological change.

THE USE-PROCESS

'The formal participation of workers in the technology use-process' (box 5, Figure 3.1) is institutionalized, via formal job descriptions, in job design. Yet trade union participation in the design of jobs with new technology tends to be restricted to issues of pay or health and safety, with the latter largely to do with hours of work on VDUs. When unions have attempted to incorporate job design issues into new technology agreements – for example, the 1985 initiative taken by the UK's APEX union (the Association of Professional, Executive, Clerical and Computer Staff) – such concerns have rarely survived the attrition of bargaining over labour redundancies and job demarcations.

Consequently, the patterns of job design that emerge out of technological change are shaped to a large extent at the organizational level. In this context, one distinct tendency, which we noted in Chapter 2, is for dominant groups to seek to maintain their power by ensuring that the application of technology maintains or increases the relative centrality and autonomy of their roles within the organization. Moreover, as psychological research (for example, Hackman and Oldham 1980) has indicated, those features of a job that we have described as enhancing its political salience also correlate to a significant degree with those features associated with job satisfaction. Two key dimensions of job design are important in both contexts. First, the extent to which the job provides the job-holder with independence and discretion: in other words, job autonomy. And second, the number of tasks that, in combination, make up the overall job: the degree of job specialization. Generally speaking, the higher the levels of autonomy and the greater the number of tasks performed, the greater the power of the job-holder.

Although conventional job design theory tends to focus on the 'enrichment' of the jobs of shop-floor workers in isolation from other tasks, the concentration of autonomy in the hands of managerial and technical groups clearly has implications for the degree of control and hence satisfaction experienced by such workers. Thus, when we map out the features of different types of job along the two dimensions cited above, we find a distribution which correlates

JOB AUTONOMY

		Low	High
		Low	High
JOB SPECIALIZATION	One or few tasks	A Routinized job (Taylorism/ Fordism) e.g. bank cashier, assembly line worker	C Specialist job e.g. systems analyst designer
	Multitasked	B Enlarged job (Flexible specialization) e.g. machine setter supervisor	D Senior management work e.g. general manager

Figure 3.2 Technology and job design

closely with the relative power and hierarchical position of different groups (see Figure 3.2).

Specialized, low-discretion jobs (box A, Figure 3.2) are the easiest to externally monitor and control. When technology is used to routinize work – as in Fordist production and Electronic Point of Sale technology in retailing – individual autonomy and discretion are diminished. As a consequence, job-holders lose bargaining power in their relationship with management because the latter's dependence on them is reduced. The thinking and planning components of work are transferred to managerial or technological control.

Multitasked, low-discretion jobs (box B, Figure 3.2) are also associated with routinizing technology but offer a greater degree of variety and skills development to the job-holder. This also has benefits for management, however; a multiskilled workforce means that workers are more interchangeable and absence through sickness or holiday can be easily covered.

Such jobs are often designed so as to accommodate the implementation and use of tightly coupled flexible manufacturing systems and production processes such as Just-In-Time. Indeed, evidence suggests that the on-the-spot problem-solving required of operators is often a functional necessity given the unreliability of state-of-the-art error detection and retrieval technology. The low levels of buffer stocks in such systems mean that failure to correct errors speedily can halt all downstream production activities.

The concept of 'quality circles' offers a prime example of the kind of formal worker participation that may be associated with low-discretion, multitasked jobs. But, Odaka (1975) argues that these are really more information-sharing and worker-involvement exercises than true participation. The 'zone briefing groups', within Rover and Honda, for instance, involve supervisors halting the assembly line in order to give information about output targets, sales performance and quality. Even in employee involvement programmes such as those employed by General Motors in the US – involving workgroup problem-solving and worker involvement in decisions about tooling and work scheduling – workers return to routinized, machine-paced tasks once collective decisions have been reached. Indeed, Wood (1986) claims that GM management intend to abandon such programmes once all production problems are 'ironed out' in the participating factories: a not uncommon managerial response to worker involvement programmes once initial problems have been solved (Dunford and McGraw 1986).

Moreover, when we apply the above analysis of job design to the concept of Flexibility, as discussed in Chapter 2, we might conclude that even here the possibilities for radical advances in job satisfaction and employee relations are limited. While proponents of Flexibility, or the related 'new production concepts' (Kern and Schumann 1985) highlight the possibilities of less specialized, 'multi-skilled' jobs, the resultant job designs do not – or perhaps, given the political interests of middle management, cannot – offer much more autonomy to job-holders than strictly Fordist arrangements would allow.

Conversely, specialized, high-discretion jobs (box C, Figure 3.2) remain very largely the domain of organizational experts. Managerial control over expertise is of a different nature to that employed for low-discretion work because of management's dependence on their expertise. Control over, rather than delegation to, experts becomes possible only if the expert's knowledge-base can be codified and routinized. We have already examined the role new technology may play in this 'de-skilling' process in some depth in Chapter 2.

Finally, multitasked, high-discretion jobs (box D, Figure 3.2) are, as we have noted, reserved for, and created by, senior management groups. This is not to say that such groups enjoy complete autonomy. Some controls and pressures do operate at this level. For instance, shareholders and suppliers of capital represent an important constituency, and senior managers are selected and socialized such that they are able to respond plausibly to the demands of that constituency. However, the logic of hierarchy colludes with the ideology of entrepreneurship in ensuring that the subjective autonomy of senior management remains relatively undimmed, even if at the expense of other groups within the organization.

'Informal user participation in the use of technology' (box 6, Figure 3.1) may be understood not only in terms of the informal working practices that arise during the use process which are crucial to effective production (Kusterer 1978, Wilkinson 1983), but also in terms of covert user attitudes and behaviour – namely, the psychological engagement and commitment to the use of technology. We find that, even in the most automated factories, managerial attempts to control the use of technology and to achieve efficiency and innovation in production depend in large part upon user engagement and commitment at an informal level (Majchrzak 1988). Thus, a study of two highly automated oil refineries, one in the UK and one in France, found that even with similar levels of technology there were significant differences in both worker productivity and management's ability to introduce technological innovation (Gallie 1978). Although differences in labour productivity may not be great in absolute terms, the competitive process has the effect of magnifying the importance of even the most marginal advantage gained from employee commitment in the production process (Sayer 1986).

Ultimately, of course, technology will not function at all without some degree of participation and co-operation from the users. Indeed, the extent of management's dependency on worker co-operation is illustrated most acutely when it is withdrawn: for example, when workers 'work to rule' or find ways of sabotaging the system. In other cases, managers' dependency on worker co-operation is only revealed when new technology is actually being used, when an ambitious blueprint is outflanked by the uncertainties and dependencies created by workers' formal and informal roles in production. This can create highly embarrassing outcomes. For example, one UK car company introduced a new machine – and worker – monitoring system into its engine production area, against union

protests and without negotiation or consultation. But having introduced it, management were surprised to discover how much the
new system depended on the accurate and timely input of data from
the workers themselves. This was, of course, not forthcoming, and a
frustrating industrial relations stalemate ensued (Scarbrough and
Moran 1986).

SUMMARY: POWER AND THE TECHNOLOGY PROCESS

In this power perspective on the technology process, we have
emphasized the extent to which the circuits of power that shape
the technology process are not predetermined in their effect but may
encompass a variety of different groups. Although the hierarchical
relations of power represented by the management–worker divide
are clearly important, we have also stressed the role played by
powerful models – notably Fordism – and knowledges in shaping
the process and its political outcomes.

For the most part, however, the analysis has centred on groups
conscious of their interests and possessed of a well-established
identity. The political manoeuvrings of such groups seem to be
explicable in terms of the kind of rational strategy exemplified by the
'power loop'. However, the construction of 'rational' political strategies also depends upon the meanings that are given to and arise
from the engagement with technology. As we will see in the next
section, the technology process may actually help to shape the structures of meaning within which groups operate (Foucault 1977), and
hence the identities and perceived interests on which their political
strategies rest.

Part II

Meaning

Power to do only resides in a context of power to define
S. Hill, *The Tragedy of Technology*

At first sight, it may seem paradoxical to discuss the 'meaning' of technology. After all, technology presents itself as an objective and tangible reality, not requiring interpretation. Machines 'seem in human eyes to act for themselves and out of their own inner necessities' (Braverman 1974: 230). However, as we noted in Chapter 1, the objectivity of the material form of technology actually disguises levels of subjective intent and knowledge. Sometimes, indeed, that objective disguise may be an integral part of achieving the intent. Shaiken, for example, describes the introduction of a factory management system: 'Had a system that time studies skilled workers been introduced as "new work rules", there might have been total uproar, but embedding it in "technology" mystifies the social content and makes it appear inevitable' (Shaiken 1980: 49).

This demonstrates the influence that the meanings attached to technology may exert on the behaviour of particular groups. It also illustrates the incentive for groups within organizations, particularly management, to exploit the symbolic value of technology as part of their *management of meaning*. This is a process of symbol construction 'designed to create legitimacy for one's own ideas, actions and demands and to delegitimate the demands of one's opponents' (Pettigrew 1987: 659).

The ability to manipulate meanings in this way, and the response to such manipulation, is likely to vary according to the roles and knowledges that different groups exercise in the technology process. To the designer or manager, new technologies may mean a challenge or a choice. To the worker, they may be presented as a product of scientific progress, or, more candidly, as simply a necessary feature of the production process.

It follows from this that ideas and perceptions of technology do not emerge automatically, but are shaped by political interests. Those groups who enjoy powerful positions within the technology process are best equipped to translate their interests and perceptions into coherent and powerful 'ideologies'. The latter are important not simply for the social currency and acceptability they achieve, but, more importantly, for the way in which they feed back into processes of technological development. Indeed, rhetorical and symbolic references to technology are visible and contestable, but the embedding of such ideologies and values within the technology process itself renders them invisible, and hence an even more powerful influence upon the lives and perceptions of all those who participate in that process. In short, the influence of technology and its ideological underpinnings is greatest when we 'live it' rather than use it. One reflection of this is the power of technology to shape meanings, as discussed below under the heading of 'the mirror problem'.

The mirror problem

The mirror problem is encapsulated in the observation that when we look in the mirror we do not see the mirror, we see 'reality'. The relevance of this observation to the study of technology is manifold.

First, it has to do with the power of technology to change our view of the world. Mumford (1934) describes how the development of high quality glass from the fourteenth century onwards encouraged a range of technological innovations – from microscopes and telescopes to mirrors. The latter helped to change the way in which humankind viewed both the world and themselves. As Mumford says, 'Glass helped put the world in a frame: it made it possible to see certain elements of reality more clearly: and it focussed attention on a sharply defined field – namely, that which was bounded by the frame' (Mumford 1934: 125–6).

And Hill (1988) extends the metaphor to technology in general, when he notes how in 'utilizing' technological objects and systems we are simultaneously creating a 'technological frame' against which all our actions whether technical or not are judged. In other words, the technology process not only frames our perceptions but in doing so shapes our values.

A simple example of such enframement might be the symbiotic relationship between the increasing use and accuracy of personal weighing machines, such as bathroom scales, and the diffusion and increasingly 'scientific' claims of diet plans and weight loss techniques.

It may take some considerable time, of course, before a particular technology becomes so 'woven into the texture of everyday existence' (Winner 1986: 12) that it becomes indivisible from social life in general. Zuboff (1988), for instance, cites an historical example: resistance to the use of written, rather than first person, evidence in medieval law-courts. She contrasts medieval suspicion of 'external marks' on a document with our taken-for-granted sense of the unity of the written symbol and the external world.

The extent to which certain technologies have achieved this kind of implicit acceptance such that we no longer 'see' their 'external marks' on social life is well illustrated by Craig Raine's (1979) poem 'A Martian Sends a Postcard Home'. By describing the strangeness of certain everyday technologies when seen through the eyes of a Martian – someone who does not take them for granted – Raine jolts the reader into awareness of his/her own unthinking acceptance. Here are a few lines from the poem:

Caxtons are mechanical birds with many wings
and some are treasured for their markings –

they cause the eyes to melt
or the body to shriek without pain.

I have never seen one fly, but
sometimes they perch on the hand.

... In homes, a haunted apparatus sleeps,
that snores when you pick it up.

If the ghost cries, they carry it
to their lips and soothe it to sleep

with sounds. And yet, they wake it up
deliberately, by tickling with a finger.

Thus, the meaning of technology does not simply reflect and reinforce powerful interests acting within the technology process, but is shaped in a fundamental way by the interpretative 'frame' around this process. Yet, as we shall see in the following chapters, the assumptive and taken-for-granted nature of this 'frame' tends to disguise the ideologies that have become embedded within it over time.

Chapter 4

Technology and ideology

INTRODUCTION

The relationship between technology and organization cannot be studied in isolation from the meaning of technology. This derives both from pre-existing social relations and the various positions that emerge from the technology process. In our discussion of meaning in this section, we employ the concept of 'circuits of meaning' to examine the technology process from an interpretative perspective. Later we look in detail at the perceptions and meanings of technology held by managers, designers and users, and the ways in which these perceptions shape, and are shaped by, the technology process. In so doing, we indicate how a wider socio-cultural ideology enframes the technology process and how this shapes perceptions of skill and identity articulated by actors within this process. For example, skill tends to be defined in relation to the job rather than the job-holder and the power to define one's work and work identity is restricted by the meanings that can be constructed within the technological 'frame'.

We begin our analysis by examining how the modern meaning of technology has emerged out of the cultural and ideological under-pinnings of industrialization. The meaning of modern technology cannot be found in the artefacts themselves but in relating the symbolism of the artefacts to the cultural and ideological context within which their meaning is located. We argue below that ideologies underpinning developments in technology during the Industrial Revolution are still very much alive today and for this reason we take time to examine the genesis and historical roots of the modern day experience and meaning of technology.

In sketching out the origins and development of the *ideology of industrialization*, we show how 'technological choice' is not solely a matter of social and political negotiation between actors or social groups holding different positions of relative power. The key to unlocking the general shaping of the technology process is the ideological 'frame' that lies beneath these negotiations.

IDEOLOGY

Ideology is a structured system of representations of aspects of reality reflected in, and reinforced by, prevailing socio-historical conditions (Albury and Schwartz 1982). Three aspects of this rather complex definition need elaborating here – not least the phrase 'aspects of reality'.

Ideologies are not necessarily true or false but are partial views of the world. Ideologies are partial in the sense that they are limited and express the view of one particular social group or coalition. A second aspect of this definition is the notion of a 'structured system'. Ideologies are coherent outlooks of particular social groups who share broadly similar material interests. Hence it is possible to speak of socialist ideology, scientific ideology, military ideology, and so forth. A third aspect of this definition relates to the ways in which ideology both reflects and reinforces prevailing social relations. Ideology often serves to legitimate the actions of powerful interest groups. Finally, ideology can be expressed in more than just words. The design of machines and buildings can powerfully embody a set of ideas. For example, heavy industry had an almost mystical significance after the revolution in the Soviet Union, a significance that disguised the exploitative role of the production technology being employed. Technology was part of the Communist ideology of the ruling Bolshevik élite, and was seen as a critical element in that group's aims. Lenin memorably summarized this intertwining of technology and ideology when he proclaimed: 'Communism is Soviet power plus the electrification of the whole country.'

The relationship between ideology and the technology process is expressed diagrammatically in Figure 4.1 in terms of 'circuits of meaning'. Ideology filters and shapes the way in which people perceive, understand and relate to a given technology, and it is the meanings attached to technology that define the technically possible and the technically desirable. Thus, the experience of technology directly influences its development – although the extent to which this influence is realized will depend on the formal and

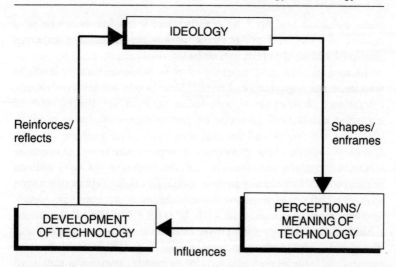

Figure 4.1 Circuits of meaning

informal power positions of the social actors and groups in the technology process. As we have seen in our earlier discussions on power, the technology process will tend to reflect and reinforce the dominant ideologies articulated by the dominant group or coalition, thereby lending support to the perceived legitimacy of pre-existing social and power relations. The example of 'efficiency' discussed in the previous chapter offers a clear example of this dynamic.

But ideology is not solely an instrument or a resource for the exercise of what Lukes (1974) terms 'third-dimensional' power (i.e. the power of one agency to manage the interpretations of reality of others). It also defines and redefines the nature and extent of the material interests it reflects as additional 'truths' about reality are revealed through the application of technology. Hence, ideologies embodied in technology not only serve to define the 'legitimate' role and status of the user, but also the role and status of those who control the technology. As modern technology increases in scale, complexity and scope, ideologies embodied within it realign accordingly.

Technology shapes our definition of 'reality' in the sense that the experience of making and transforming the world is mediated by our perceptions of what technology makes possible. Even our definitions of 'life' and 'death' have been altered by modern technology – medical technologies such as life support machines, internal organ transplantation and artificial organs make it possible for those who control the technology to treat and cure medical conditions previously defined as terminal or fatal.

Ideologies exist at a macro – that is, socio-cultural – level, as well as at the organizational level. These two levels are inextricably interrelated. For example, employees tend to see the exercise of executive managerial power as legitimate only in so far as it aligns with wider cultural and societal ideologies. The exercise of what Lukes (1974) terms 'one-dimensional' power – the power of coercion – cannot explain the growth and development of the modern organization: 'Legitimacy persists as long as daily experience aligns with perceptions of meaning (and horizons of accepted privilege), and trajectory of the order as a whole' (Hill 1988: 104). If this alignment breaks down, existing power interests become transparent and illegitimate, and challenged accordingly. As we shall see in the next chapter, so long as technology is perceived by users as a means to achieve a higher standard of living and to symbolize progress, its ideological frame will tend to remain hidden and taken for granted, thereby perpetuating the prevalent circuits of meaning. Thus, while power and ideology are interrelated, one cannot be subsumed conceptually under the other.

SOCIO-CULTURAL IDEOLOGY

The ideology of industrialization

In Europe and beyond, the nineteenth century was a period of rapid change and conventional wisdom gradually swung round to the idea that, contrary to traditional religious beliefs, the world really was changing. The economics of wealth creation became a key to the development and progression of nineteenth-century England and industrial technology was destined to play an important role in this process with labour productivity as the driving force. In essence, the argument – epitomized by economists such as Charles Babbage (1832) and Andrew Ure (1835) – stressed that economic growth was not feasible without increased labour productivity. In short, individuals within a society must produce more than they consume. For both Babbage and Ure, the subdivision of the labour process, coupled with the development of faster production machinery, provided the foundations upon which labour productivity (and thus societal wealth creation) would be built.

Hence, the development of technology in the early nineteenth century should be seen in the context of an ideology which equated social and economic progress with labour productivity. Both Babbage and Ure realized the profitability inherent in keeping labour costs low and emphasized the key role machine technology could play in this. Ure, in particular, was aware of the need to maintain a strict discipline in the organization of a workforce if the benefits of increased labour productivity were to be reaped, and he saw developments in technology as providing an excellent means of achieving this. His insight was no doubt informed by the fact that the British government had felt it necessary to deploy some 12,000 troops between 1811 and 1813 to quell outbreaks of machine breaking by disgruntled workers.

Confrontation between owners of factories and their workforces became increasingly frequent after the 1820s. Machines began to be introduced not only to aid the enforcement of discipline but also to undermine growing worker militancy. Dickson notes that 'these tactics inevitably included the need for increased social control on the part of capital, and the authoritarian relationships that this implied became crystallised in the machines that were introduced' (Dickson 1974: 79).

Of course, the developments in production technology (and the social relations they reflected and reinforced) meant something quite different to the employer and the worker. To the worker it meant less control, an erosion of craft skill and, frequently, lower wages. To the employer, technology meant higher productivity and increased control over (and enhanced predictability of) the work process. This conflict of interests and values cannot be resolved by deciding which view is right and which is false – they are both partial views linked to different material interests. As such they are clearly ideological. Yet, the capitalists' view was all the more influential because of the wider ideology of industrialization which equated higher labour productivity with increased economic prosperity for all. Indeed, the growing prosperity in evidence in England at the time of the 1851 Great Exhibition further legitimated this interpretation of reality. To reject industrialization was to reject economic prosperity and return to the relative poverty of feudal society. In this socio-cultural context it is small wonder that the Luddites received such a bad press!

Industrialization, science and technology

The development of technology after the Industrial Revolution in the second half of the nineteenth century was deeply rooted in the development of scientific knowledge and its application to problems of social and economic development. Until this time, technology had been designed and implemented predominantly by practical people in the organizations within which they worked. In the latter half of the nineteenth century, scientific rationality began to take possession of domains of experience far beyond mathematics and the physical sciences. Mumford makes the point that major technological innovations now derived from initiatives of scientists rather than indigenous inventors. This gave rise to a new design phenomenon – deliberate and systematic invention: 'Here was a new material: problem – find a new use for it. Or here was a necessary utility: problem – find the theoretical formula which would permit it to be produced' (Mumford 1934: 218). The extension of a scientific interpretation of 'reality' into the social world has had a profound impact on the technology process ever since. The reason for this stems from the social ramifications of scientific thinking and logic, focusing on observation, explanation, prediction and control.

Stanley (1978) argues that the encroachment of scientific rationality into the socio-cultural world has had two major consequences. First, it has led people to ignore the essential differences and discontinuities between the human social world and the worlds of mathematics and the physical sciences. As we shall see, the ideology of scientific rationality encouraged the view (among employers and managers) of work organizations as giant mechanisms or machines to which scientific principles and laws could and should be applied.

A second consequence of extending scientific rationality into the socio-cultural world is that it has led to a belief that any mode of thinking and acting that is not scientific is somehow inferior to thought and action informed by the constitutive assumptions of science. In work organizations, for example, management will often quote 'efficiency' as a key objective for technological innovation. Efficiency is then defined relative to the quantifiable attributes of technology. In not lending itself to simple measurement, labour invariably figures as an 'inefficient' variable in management equations. By this logic, labour efficiency increases the more it is predictable, routinized and/or controlled by technology (i.e. more machine-like). To argue against the efficiency of this routinization process requires 'stepping outside' the ideological frame.

Of course, human dignity, freedom, purposive action, creativity, unpredictability and intuition are beyond the understanding of such scientific rationality. This paradox is neatly summarized by Hampden-Turner who points out that

> if the social sciences [such as management science] are to build a predictive theory of high precision and invariability then healthy and creative personalities are a major snag, while the sick, the incarcerated and the conforming will be in great demand.
>
> (Hampden-Turner 1970: 9)

We shall look at the ramifications of this in Chapter 5.

Hill makes the important point that the extension of scientific rationality into organizational ideology and practice follows from the logic inherent in scientific modes of thinking. In assessing the development of industrial technology, he opines that:

> the logic of science offered a way of comprehending and creating systems of technical action. The aesthetic implication of this logic followed, that is, that the systems should smoothly interconnect, and inefficient or dislocating components within the systems should be eliminated. This observation leads directly to a fundamental difference between the machines of the industrial age and those that went before. For the inefficient component of a technological system lies not so much in the technical arrangements as in the human labour that connects them.
>
> (Hill 1988: 50)

Hence, the application of technology to overcome worker militancy prior to the Industrial Revolution can be seen as the exercise of coercive power by employers in their efforts to discipline and control labour. But, with the social and economic benefits spawned by the invention and widespread application of electricity, a new ideology emerged which promoted and reflected the belief that all social and economic problems were capable of rational solution through the discovery and application of scientific principles. The application of technology could still be (and was) employed with the explicit intent to control labour, but the new ideology served to obscure the exercise of 'one-dimensional' power.

This socio-cultural ideology – embracing the systems logic of science – proved to be fundamental to both the technology process and to organizations in general in the twentieth century, as we outline below.

ORGANIZATIONAL IDEOLOGY

The systems logic of science and organizational strategies of labour control

Miller and O'Leary (1987) argue that by the beginning of the twentieth century, when both engineering management techniques (particularly scientific management) and standard cost accounting were being developed in earnest, the notion of 'calculability' (the use of accounting records as a means of predictive control) became increasingly important within management thinking. Taylorism may be seen as a clear example of this notion couched in engineering terms. For Taylor, the efficiency of an employee cannot be observed with the naked eye and cannot be measured until what is to be regarded as normal or standard has been constructed.

But, as Ovitt points out: 'Until there is a cultural consensus supporting notions of productivity and efficiency for their own sake, there is no incentive within a social system – any social system – to concentrate on developing techniques for the tedium of labour' (Ovitt 1986: 500).

Taylor's 'scientific management' was to become so influential because it held out the promise of high productivity (and hence increased economic prosperity) and industrial democracy through its alignment with the wider socio-cultural ideology of industrialization and the systems logic of scientific rationality. As Rose points out:

> Taylor's technical proposals – his system – was, then, a set of devices which sought to capitalise on workers' rationality and suppress managerial irrationality. It would eradicate inefficiency and arbitrary managerial prerogatives (and thus the workers' rebellious response to them). In sum, it would create an authentic industrial partnership. But why should either party accept it? And who would form the vanguard which could secure its adoption? Taylor's reply to the first question was, because it was *scientific*: impartial, universal, *lawlike* in both the scientific and judicial senses. To the second his response was, the production engineer.
>
> (Rose 1975: 34)

Rose observes that if Taylor had been a conscious fabricator of social myths, he could hardly have done better than to christen his system 'science'. This instantly appealed to the new middle-men of industry. In ideological terms, science supplies the 'one best way' of producing goods for all. Within its own frame of reference it is neutral, progressive and able to provide solutions to the problems of labour control and co-ordination (as well as generate high profits, high wages and supply the increasing demand for technological goods). Indeed, Taylor confidently expected his system to greatly improve the relationship between management and workers through their mutual submission to the 'neutrality' of his scientific methods and techniques. It came as quite a shock to him that management and workers alike responded, at least initially, with great hostility to the implementation of his system.

The main reason for this hostility stemmed from the assumptions inherent in scientific management. Certainly productivity could be greatly increased through the adoption of the 'one best way'. But this increase was at the expense of employee self-direction and expertise, as well as posing a threat to the power of non-engineering management (see Chapter 2).

Just as scientific rationality encourages the view of scientists as the custodians of 'truth', Taylor's 'scientific management' encouraged the development of managers as the legitimate holders of explicit knowledge of the organizations in which they served. Thus the management function became one of rational planning and control, while the role of the workforce was to follow these plans (thereby relinquishing control over their working methods to management). With their superior overall knowledge of the organization, management's role became that of uncovering the optimal method (or 'one best way') of working, while the role of the workforce was to follow the rules and procedures so derived. Although this process of rationalization effected a substantial shift in power, the fact that it went hand in hand with perceived increases in productivity and more secure employment for the workers precluded any large-scale opposition to its encroachment. Again we see the inherent difficulties in breaking the 'circuits of meaning' around the development of technology.

Taylor's system supplied both a justification and a means for the expropriation of craft skills and knowledge, and the control of labour. But it was the design and implementation of the assembly line by Henry Ford at his vehicle plant in Highland Park, Detroit, in 1913 that revealed how Taylor's principles could be extended into the design of physical machinery. The technology of the assembly line enabled the imposition of a centralized machine pacing of parts transfer and short task cycle times. Instead of workers moving at their own pace between tasks, the tasks move to the workers at a pace determined by the technology. The technology greatly increased the predictability and controllability of the labour process, thus paving the way for additional technological 'improvements'.

This revolutionary mode of organizing and controlling production was firmly established in the US by the end of the 1920s and spread (to a lesser extent) to Europe, Japan and elsewhere. It proved to be so successful that it led to rapid increases in productivity and increases in employment and wage levels, albeit with significant regional variations. As with the factory system and the implementation of Taylorism, Ford's assembly line provided material support and legitimation for the application of scientific rationality in organizations. In France, even the militant CGTU Trade Union leadership could find no reason to oppose the introduction of the assembly line. The line was 'a finding of science. You cannot fight against science or machinery any more than you can fight against the rain' (Fridenson 1978: 169).

Despite the pioneering work of Henry Ford, the mass production assembly line was not to become the dominant production system design model in the industrialized nations. A majority of manufacturing at this time was (and still is) carried out using batch production techniques. In the accountant's calculus, traditional batch production costs approximately twenty times more than mass production of an item owing to the delays involved in continually resetting machines and in transferring a variety of components between machines. Most items spend long periods of time sitting on the shop-floor queuing for the next stage of the production process.

It was the new discipline of Operations Research (OR), peopled by ex-military scientists and engineers and inspired by developments in automatic control technologies, that offered the opportunity to rationalize the relatively expensive and unpredictable batch production system through the ever more sophisticated expansion and control of productive time.

By the end of the [second world] war there emerged a theory of servomechanisms that was universally applicable and easy to manipulate. Moreover, there was now a mature technology of automatic control, which included precision servomotors, for the careful control of motion; pulse generators, to convey electrical information; transducers, for converting information about distance, heat, speed, and the like into electrical signals; and a whole range of associated actuating, control, and sensing devices. Finally, the wartime research projects had created a cadre of scientists and engineers knowledgeable in the new theory of servomechanisms, experienced in the practical application of such systems, and eager to spread the word and put their new expertise to use.

(Noble 1984: 48–9)

The 'word' was that the rationalization of the batch production process and the control of labour could be achieved by viewing the batch manufacturing factory as a closed, mathematically definable system. OR systems models allowed an entire production process to be reduced to discrete components whose interrelationships could be determined mathematically and then reassembled into a more predictable and formally controllable system.

From the late 1950s on, the concept of the 'system' came to dominate the thinking of both engineering designers and management to an unprecedented degree (Boguslaw 1965) and, by the 1960s, systems engineering and systems analysis were eagerly applying the systems logic of scientific rationality to a wide range of organizations – industrial, retail, service, military and governmental.

Scientific rationality and the objectification of time

Thompson (1967) argues that the control of time provides a key element in the application of scientific rationality to the control of labour. The ideological character of this process is historically rooted in the rise of industrial capitalism which brought with it a linear conception of time closely equated with the normalizing judgement of value. Technological innovations saw the concept of time become closely aligned with that of industrial and economic progress to the extent that time became objectified into a valuable commodity of the production process.

In this light, Mumford (1934) emphasizes how the clock rather than the steam engine may be seen as the key technology of western industrialization. He argues that rapid developments in synchronization were responsible for organizations of the Industrial Revolution being able to display high levels of functional specialization.

Intraorganizational co-ordination required planning, and sophisticated temporal schedules became necessary to provide a degree of predictability. Hence the time period replaced the task as the focal unit of production and the clock became a crucial technology for co-ordination and control within organizations.

For Taylor it was the job of scientists to determine standard times for particular jobs so that production technology could be run at the appropriate speed and workers would get a fixed wage for meeting the 'norm' plus a bonus for faster work. Taylorism epitomizes an ideological view of organizations in which

> Ideal organisations are those having temporal assets which are highly precise in their structuring and distribution. As technological determinism dominates our perceptions of time, so correct arithmetical equations are seen as the solutions to time problems; there are finite limits and optimal solutions to temporal structuring.
>
> (Hassard 1989: 19)

This view, of course, is still with us today. Consider the way time is constructed in present-day organizations. Grossin (1969) argues that the commodification of time and the gearing of entire organizations to the avoidance of loss of time ('lost time is a waste of money') is symptomatic of the encroachment of scientific rationality into the social realm. Management often legitimate the exclusion of users in the technology process by arguing that it 'slows down' implementation, and the current fascination with Just-In-Time production techniques equates time with speed of productivity, inventory and work buffer levels, and responsiveness to market demand. Indeed, the control of time, and hence labour, lies at the heart of all current production strategies – whether Fordist or Flexible.

It is in the technological realm of objectified time that the relationship between power and ideology may be delineated more clearly. Historically, professional and managerial work has tended to be task- rather than time-oriented: less bound by the 'tyranny of the clock'. But even the relative autonomy enjoyed by these more powerful organizational groups is now under threat as computer technology enters the office environment (see case example of the 'man-month'). Within the ideological frame of the technology process, middle management resistance to 'chronarchy' is likely to be as unsuccessful as the resistance put up by skilled craftspeople many years earlier.

Case example: The 'man-month' [sic] *or how to accelerate the human machine*

Within 'scientifically rational' organizations, employees themselves come to be treated as machines. Employees are required to conform to a highly structured, formal and quantitative definition of work time. Chronarchy, the rule of time, results in the extensive managerial control of time. But this rule of time is now extending into the 'white-collar' domain.

Take, for example, the contemporary notion of the 'man-month' or 'man-year'. Managers of project or task teams, forever trying to meet deadlines, often estimate tasks in such terms. This rigid system inevitably leads to problems, for it suggests that what ten people can do in four months (forty 'man-months'), forty people can do in one month (forty 'man-months'). Such a way of thinking assumes that human effort can be accelerated simply by adding more human machines – regardless of the nature of the task being undertaken and the management problems generated as a result.

This book took approximately two person-years to write. Thankfully, the publishers did not come up with the idea of contracting twenty-four people to write it in the space of one month!

The systems logic of science and the strategic control of organizations

It is not just the internal workings of an organization that are enframed by ideology. How organizational decision-makers perceive the environment shapes managers' perceptions of the market and defines what actions are possible and potentially effective. Information from the market environment is 'understood' through an ideological filter which biases managers towards giving attention to some aspects of the market and ignoring others (Thompson and Wildavsky 1986). Thus, senior managers effectively create an agreed model of the environment which determines the nature of the world they actually perceive. The systems logic of science has informed this process of 'environmental enactment' (Weick 1969) since at least the early years of Fordism.

Henry Ford realized the importance of applying technology to the control of both production and the consumer market. Mass-production technology requires a mass market and Ford looked to standardize marketing, sales and service through what Veblen (1923) called the 'quantity-production of customers'. Ford achieved this through advertising (using the new radio and newspaper technology) and viewing the workforce as customers (the famous five dollars a day income was significantly above basic subsistence levels).

Like Taylor, Ford viewed the organization as a large complex machine. As such, an organization must be designed so that uncertainty is eliminated as far as possible and organizational behaviour is constrained and controlled to achieve the desired outcomes – the goal of OR. But, it also follows that the organizational environment must be stable enough to ensure that the products produced by the 'machine' are appropriate ones. Certainly Alfred Sloan, President of General Motors, realized that markets were a prime area of uncertainty. An organization must attempt to reduce the autonomous character of the demand for its products and increase its induced character. One way to achieve such market stability is to gear customer 'needs' to the 'needs' of production instead of the other way round (Sloan 1954).

Hill (1988) argues that the control of markets is a natural property of the general trajectory of technological systems development:

A system involves an ordered relationship between ordered parts, and closure from an unordered environment from which necessary system inputs are drawn in a strictly uniform or ordered form that can be incorporated into the internal order of the system. Thus, the most efficient systems are those that can draw from an 'environment' that is itself in ordered alignment with the standardised requirements of the system itself.

(Hill 1988: 53)

Three social inventions had to be put in place for the purchasing ability and desire of consumers to be brought into line with the technical system that produced them. These were (a) the invention of consumer credit, (b) the invention of obsolescence built into commodities, and (c) the invention of 'marginal differentiation' of commodities so that their meaning to the purchaser could reflect individual social status while production remains on mass-production principles.

The American culture of the time increasingly placed emphasis on individuality rather than on collective consumption. Yet the application of scientific rationality to production systems and product design emphasized standardization and uniformity. The third social invention, of 'marginal differentiation', solved this paradox by ensuring that the core of the production process remains standardized, but with minor cosmetic product variations (e.g. body colour, body trimmings and internal fixtures) being introduced towards the end of the production process. Such practices are far more widespread today in both Fordist and Flexible organizations and reveal how the technological 'frame' extends from organizations into the wider world of the consumer.

New technology and post-Fordism

For a number of writers, the development of the digital computer and the microprocessor marks a change in the general trajectory of technological development, heralding a 'second industrial revolution' (Forester 1980, Toffler 1980). Computer technology is viewed by researchers within the Choice camp (see Chapter 1) as an enabling technology allowing organizational decision-makers to pursue new strategic initiatives inside and outside the organization. In this view, computer technology opens up new opportunities, possibilities and areas of strategic choice. Production can be fully automated or redesigned to enable group working. Redundancy, de-skilling and skill upgrading are all possible – it is all a matter of organizational choice.

The reasoning behind this claim stems from the technical capability of digital computers and microprocessors to automatically monitor and control previously manual-intensive work processes and to rapidly process vast quantities of data to and from all organizational levels and functions. Hence the 1960s and 1970s heralded the outpouring of countless futurist scenarios – the peopleless factory, the paperless (perhaps peopleless) office, the end of ideology, the end of bureaucracy, even the advent of true participatory democracy.

However, many writers argue that the very success of the mass-production/mass-consumption system has led to a degree of domestic market saturation that fundamentally questions the ideology of scientific rationality upon which the system is based (e.g. Kern and Schumann 1985, Piore and Sabel 1984). Far from seeing computer technology as a determinant of this trend, these writers point to the impact of macro-economic factors. They opine that the market saturation brought about by the successful application of Taylorist and Fordist principles can be overcome only when industrial organizations reject the 'predict and control' approach to organizational behaviour. In this view, predict and control, an essential tenet of scientific rationality, needs to give way to an approach based on 'softer' principles of responsiveness, learning and adaptation (Ackoff 1983). This would enable organizations to respond in a 'flexible' way to new market opportunities and gain new market niches through a strategy of Flexibility (see Chapter 2).

The implications for strategic choice are far-reaching. Such a reformation of the 'predict and control' basis of scientific rationality would entail a corresponding change in the social relations of production. Because of this political dimension, and particularly the resistance that tends to be generated by those groups whose power base is threatened by such a radical change, Flexibility remains a relatively powerless ideology. Hence, rather than seeing what Kern and Schumann (1985) call a 'reprofessionalisation' of the productive labour process, we see many organizations (particularly in the chemical and steel industries) responding to market saturation with a strategy of innovation and production specialization while keeping intact the Fordist apparatus of production and work organization, industrial relations and skill restriction (Matthews 1989).

Indeed, no major change in the general trajectory of technology is apparent. Computer technology has facilitated a degree of convergence of mass and batch production principles and techniques in the form of flexible manufacturing systems (FMS) and computer integrated manufacturing (CIM) systems. It is now possible to automate even under conditions of market fluctuations and unpredictability (Littler 1983).

The reasons for this are certainly political, but also ideological. As an ideology, Flexibility differs from Fordism far less than its advocates generally claim. Both can be seen as expressions of the ideology of industrialization (stressing the importance of disciplined labour productivity) and the systems logic of science (emphasizing standardization, predictability and control). They have in common the precise technical control of productive labour time, for example. Hence, at the level of shop-floor job design, there is little to distinguish the 'flexible specialisation' of Flexibility from the routinized specialization of Fordism (see Chapter 3). Labour 'efficiency' – whether in terms of work quantity or quality – is judged according to standardized technical criteria in both systems. Ultimately,

Fordism and Flexibility employ the fundamental ideological principle of ordering social relations of production to align with the technical order.

SUMMARY: POWER, IDEOLOGY AND THE TECHNOLOGY PROCESS

Science and technology are not neutral – in a political sense – on two counts. First, when applied to social reality, scientific rationality is intrinsically conservative. The need to predict and control produces detachment and an unequal relationship between the observer and the observed, the controller and the controlled, the expert and the user. It emphasizes the *status quo* above visions of future possibilities and opportunities. Also, scientific rationality effectively banishes human purpose from its discourse and reduces the person to the status of an object. Ultimately, of course, science is deterministic as our behaviour is seen as being reducible to, and determined by, universal scientific laws and principles. Behaviour that does not conform to these laws, for example, 'inefficient labour', can then be easily branded as 'deviant'.

Second, the brief history of technological development outlined above reveals that the technology process can express and reinforce the interests of the employers and managers of work organizations. This is not to suggest that technology is in some way a direct expression of the purposes of one distinct interest group. As we have stressed throughout the book, the multifaceted nature of technology makes such control extremely difficult to achieve. Ellul (1974) argues that the encroachment of scientific rationality into social and work life is not the product of some conspiracy on the part of managers, engineers or 'technocrats'. Rather, people are *drifting* into a subordination to the ideology of scientific rationality, to the conservative values and rationality of science and its enframement of the technology process. Our discussion of power in Part I of this book would suggest that, in addition to this general *drift*, scientific rationality remains a highly influential ideology in technological and organizational design because it offers, simultaneously, techniques which control uncertainty (and therefore grants power to those who can apply it in the managerial interest) *and* a legitimating frame that *appears* free of political bias.

Chapter 5

The meaning of technology in organizations

INTRODUCTION

The meaning of technology in organizations is fundamentally linked, via circuits of meaning, to the relationship between subjectivity and objectivity. The subjective elements within the technology–organization relation include the perceptions of designers, managers and users; the objective elements include the ways in which subjective knowledge becomes objectified in technology, in job design, and in the expertise and expert power derived from the control of technology. Rosenberg regards this process of objectification as a key characteristic of modern technology. He argues that modern industrial organizations have a distinctive technological feature:

> It is that, for the first time, the design of the productive process is carried out on a basis where the characteristics of the worker and his physical environment are no longer central to the organisation and the arrangement of capital. Rather, capital is being designed in accordance with a completely different logic, a logic which explicitly incorporates principles of science and engineering. The subjectivity of a technology adapted out of necessity to the capacities (or, better, the debilities) of the worker is rejected in favour of the objectivity of machinery which has been designed in accordance with its own laws and the laws of science.
>
> (Rosenberg 1976: 132)

For less powerful others, the meaning of technology is relatively opaque and they will tend to rely on the definitions of reality – that is, ideology – given by more powerful technology-shaping groups. In analysing this process from the shop-floor perspective, Burawoy comments: 'Objectification of work, if that is what we were experiencing, is very much a subjective process.... We participated in and strategized our own exploitation. That, and not the destruction of subjectivity, was what was so remarkable' (Burawoy 1985: 10).

We now turn our attention to the ways in which the socio-cultural ideologies of industrialization and systems science, outlined earlier, permeate the management of meaning of technology at the organizational level.

TECHNOLOGY AND ORGANIZATIONAL CHOICE: MANAGERIAL PERCEPTIONS

Within the Choice literature on technology and organizations (see Chapter 1), much has been made of the importance of the strategic choices available to organizational decision-makers when planning and implementing technological change (see Buchanan 1983, Sorge *et al.* 1983, Child 1987b). What remains relatively under-researched are the ways in which management (working at different levels and in different functions within an organization) perceive technology and technological change and the role that these perceptions play in the process of choosing particular technological designs and configurations and of implementing these choices. In what ways, and to what extent, can ideology enframe such a process?

Many writers argue that it would be misleading to view organizational decision-making as a rational process undertaken by an homogeneous social or political group (Mintzberg 1978, Pettigrew 1985). But, when we are told that decisions concerning technological choice are made by 'management', the heterogeneity of management and the influence of inter-professional competition, discussed in Chapter 2, tend to be overlooked. Yet such competition is a fundamental aspect of technological change and was one of the reasons for the demise of 'classical' Taylorism (see Rose 1975: 47–53). Taylorism not only changed traditional managerial roles, but tended to shift decision-making powers away from owners, financial administrators and accountants to engineering and production management.

Different levels and functions of management will tend to hold different perceptions of technology and technological change in organizations. The research of Mueller *et al.* (1986), based on extensive interviews in a large UK computer manufacturing company, clearly illustrates the heterogeneity of management views

Table 5.1 Management preferences for the control of new
manufacturing technology

Locus of control	Managerial function/role
1 Strongly prefer control by specialists	Manufacturing Engineering Management Information Systems
2 Prefer control by specialists	Design Engineering Production Technical Operations Production Planning and Control Maintenance Engineering
3 Mixture or 'not known'	Test Engineering Financial Controller Operations Director Supervision
4 Prefer control by operators	Training Quality Control Quality Engineering Personnel
5 Strongly prefer control by operators	Technical Support Stores

Source: Mueller *et al.* 1986

concerning the operation and management of advanced manufacturing technology within an organization. Table 5.1 contains a
classification of the managers' perceptions along a centralized–
decentralized control dimension.

Mueller *et al.* (1986) found that the choice between operator
control (decentralization) and specialist control (centralization) was
perceived in very different ways by different groups of managers.
However, it appears significant that those managers with the technical expertise necessary to play a prominent role in technological
implementation were decidedly not in favour of decentralized
operator control. Interviews revealed that these engineering
managers felt that reduced human intervention would lead to
benefits such as a reduction in 'handling errors'. This view reflects
engineering management's allegiance to the systems logic of science
and the design principle of 'prediction and control'.

The ideological nature of such a view is clear when one considers
that computers are viewed as a means to reduce uncertainty

whereas evidence suggests that computers (particularly software) are often the source of uncertainty (see case example below).

Case example: Computers as generators of uncertainty or why software is not guaranteed

'Suppose a programmable electronic system is monitoring 100 binary signals – either inputs or stores within its memory – to detect changes in operating conditions. This means there are 2^{100} (1.27 × 10^{30}) possible combinations of switch inputs and memory contents. In addition, the part of the program that the microprocessor executes will depend on the position of the switch, the contents of the memory, and on the functions being carried out: there may be 10,000 (10^4) possible paths through the program. As a consequence, designers would need to do 1.27 × 10^{34} individual tests on the system – taking every combination along each path – to guarantee it contained no faults. If they could do the tests automatically at a rate of 100 tests per second, they would need 4 × 10^{24} years to complete – assuming they could identify all the possible paths through the program. That is many times the life of the Universe – and this program is a relatively trivial one. Furthermore, some data paths, such as error recovery routines, may be followed only when an internal computer failure occurs. It may not be possible to test these paths without first modifying the program which, in itself, may lead to the introduction of additional faults.'

Source: Wray (1988: 61)

The dominance of this engineering logic of efficiency in the decision-making process is highlighted in a number of recent case studies of European manufacturing organizations (e.g. Kesteloot 1989, Kulpinska and Skalmierski 1989). Examining the choice and introduction of a flexible manufacturing system, Kulpinska and Skalmierski argue that managerial cadres within Polish industry are dominated by engineers who are oriented to technical problems of production and who do not possess any special qualifications in the area of production organization and management.

A basic element of their professional ideology is fascination with technology which they regard as the only remedy for production problems which in reality are mainly due to characteristics of the work system and, thus, the nature of work organization. Another element of this ideology is a desire to promote technical progress by any means as an autotelic value.

(Kulpinska and Skalmierski 1989: 143)

Such case study findings indicate that while engineers do not necessarily hold a strong position in the competition for power resources between professional managerial groups, scientific rationality enframes the decision-making of other expert groups within organizations to the extent that it aligns with their own interests and control objectives (see Chapter 2 and Morgan 1990). Within this ideological frame, the range of possible ('realistic') strategic choices is predefined.

This is evidenced by the second important aspect of Mueller *et al.*'s (1986) findings – none of the managers in their study expressed any doubts concerning the need to undertake the technological change in question, nor did they question the appropriateness of the chosen new technology itself. Both were taken as given. Hence, although there was no apparent uniformity of views concerning the main objectives of change or the appropriate type of job design associated with new technology, there was a significant consensus on the need for, and perceived desirability of, technological change. Heterogenous perceptions and intragroup and intergroup negotiations undoubtedly play a part in defining the use of technology, but this is not the whole story.

These two factors, technology as a means to reduce uncertainty and technological change as a necessary and desirable means to achieve a variety of interrelated organizational objectives, illustrate how scientific rationality enframes managerial decision-making in the technology process. While the vested interests of different managerial functions produce a heterogeneous set of beliefs concerning the use of technology, these interests reflect and reinforce the perception of the technology as beneficial and progressive. For instance, many organizational decision-makers and management consultants share a belief that is summed up in the phrase: 'automate or liquidate'. In addition, it is not uncommon for management consultants and organizational theorists to argue that organizations should be restructured in order to gain the full benefits of new technology (see discussion in Wilson 1992).

Despite (or perhaps because of) this deterministic view of technological development, case study research reveals that the management of technological change in many (perhaps a majority of) work organizations is one-sided and dominated by technical considerations. At the shop-floor level, it seems that change is achieved more by 'muddling through' the social arrangements around machinery than by any coherent strategic choice (Wilkinson 1983, Clegg and Corbett 1987, Majchrzak 1988, Armstrong 1988, Heller 1989). And, even if engineering managers themselves do not possess or exploit the power to shape the design and use of technology within their own organizations, it appears that some form of engineering logic will tend to prevail during the management of change. In effect, as the case example of NC machine tools demonstrates, we may even see technology itself becoming a symbol and supporter of this engineering systems ideology. In turn, the meanings attached to technology further reinforce acceptance of the domination of systems specialists and engineering experts in the decision-making process. It could even be argued that senior managers find it expedient to encourage the domination of technical considerations during periods of technological change in order to mask the political nature of change.

Case example: Ideology and the development of numerically controlled (NC) machine tools

The challenge of automating the cutting of metal has long been a technical preoccupation of engineers. It became realizable after the Second World War as a result of the development of magnetic tape reading instruments and more sophisticated machine controls. The challenge was the preparation of the tape. Noble (1985) demonstrates that a viable solution – 'Record-Playback' – was found at General Electric in 1947. This involved having a skilled machinist make a part while the motions of the machine were recorded analogically on tape. After the first part is made, identical parts can be reproduced by 'playing back' the tape through the metal cutting machine.

However, an alternative solution was developed at the Massachusetts Institute of Technology – numerical control (NC) – which relied on the development of a specialist programming language. This language required the use of knowledge and skills very different from those associated with the operation of the metal cutting technology of the time.

Noble shows how GE and Lockheed began with 'Record-Play-back' but quickly switched to NC, despite enormous technical problems which would never have arisen with 'Record-Playback'. He argues that the reasons for abandoning 'Record-Playback' were predominantly ideological. While 'Record-Playback' required input from a skilled machine operator, NC relied on engineers and programmers giving orders to the operator. The latter separated planning from doing, gave control of the technology to office personnel and engineering management, and enabled the operation of metal cutting machines to be undertaken by a less skilled (and therefore less expensive) workforce.

Noble argues that the reduction of operator error and unpredictability facilitated by NC technology is the engineering expression of management's attempts to minimize their dependence on labour by increasing their control over production. Hence the ideology of scientific rationality within engineering mirrors social relations of capitalist production, and this, in turn, is reflected in and reinforced by the design of NC technology.

So it is that social and ideological choices came to be embedded in machines prior to their implementation. The consequential importance of these ideologies in shaping technology leads us into a more detailed consideration of the perceptions and practices from which such ideas and 'truths' emerge.

TECHNOLOGY AND ORGANIZATIONAL CHOICE: PERCEPTIONS OF SYSTEMS DESIGNERS

The design of any technical system that involves human beings as operating parts amounts to a partial restructuring of social roles and relationships (Winner 1983). For example, embedded in any technological system is a systems architecture that defines important aspects of its operation and control in terms of task inter-dependencies and work flow. Turner and Karasek point out that the choice of systems architecture and work flow is a decision from which most other decisions about technical systems design flow: 'Whether systems designers realise it or not, the decision is tantamount to defining the human's job. Hence, decisions on this dimension are not merely technical but organizational choices' (Turner and Karasek 1984: 671).

Engineering management and systems designers tend to perceive organizations as relatively 'closed systems'. 'Closed system' thinking is characterized by tools, techniques, practices and language which embody an approach to the world as composed of interlocking systems susceptible to formal mathematical analysis. Edwards (1985, 1988), for example, sees the intellectual heritage of systems science to be mathematical games theory, linear programming, operations research and systems analysis. These theories focus on the relationships between elements of a system, rather than isolating the elements themselves for individual study. Since they are essentially engineering approaches designed to solve real-world problems, systems theories tend to assume the closure of the systems they analyze. Hence, Lilienfeld argues that, in this guise, systems theory finds its chief application as the 'ideology of bureaucratic planners and centralizers' (Lilienfeld 1978: 263).

This is not to suggest that systems do not grow. On the contrary, as systems of technical knowledge expand they enable additional aspects of reality to be incorporated into this 'closed world'.

Research on the cognitive styles and design work of operations research and systems analysis professionals presents a clear picture of the ramifications of a 'closed' view of reality (DeMarco 1979, Keen and Bronsema 1981, Corbett et al. 1991). For example, in her review of the research, Markus (1984) argues that the thinking and problem-solving styles of systems analysts differ from those of systems users and yet their system designs are often built on the implicit assumption that they are the same. Analysts fail to see the political nature of organizations into which their systems will be placed and assume that all problems have a technical solution. Systems professionals also make assumptions about employee work practices and needs that tend to be inaccurate and yet are reflected in designs that push jobs into being well defined and structured, with carefully set targets and close supervision (see the DVLC example in Chapter 6). Finally, systems professionals emphasize the need for the latest hardware and software even where no clear benefits are likely to accrue to the user organization from their acquisition. This may also be interpreted in political terms as the effort to expand their expert knowledge-base and power.

Pacey (1983) argues that this restricted style of thinking – what he terms 'professional misperception' – is not motivated entirely by political interests, but is related to the 'closed world' intellectual culture of technology and the habit of identifying the use of technology with its strictly technical aspects. Systems analysts show a consistent tendency to 'see' only those aspects of technology use that are of direct technical interest. For example, there is a tendency among technical experts to give the label 'side-effect' (rather than 'effect') to any unanticipated human or organizational effect caused by technological change. Systems analysts reduce the world to an 'expert sphere' which they know in detail, leaving a completely different perspective – a 'user sphere' – which they either ignore or (incorrectly) assume to be commensurable with their own sphere of knowledge. As a result of this disjuncture

> systems theory, usually worked up with the aid of the computer, sometimes serves only to lend a bogus air of precision to a basically imprecise approach. It even obscures the qualitative insights which are the chief value of the systems approach, and it provides experts with new techniques for mystifying and manipulating users.
>
> (Pacey 1983: 51)

The 'bogus air of precision' surrounding systems theory is shared by computer systems specialists and is often reflected in the language they employ. Consider the concept of the computer program 'bug'. Bloomfield points out that, as well as meaning a defect, 'bug' is also synonymous with the idea of something small and insignificant.

> The problem here is that while many program bugs may well be insignificant there may exist others which, depending on the particular application, may be more or less catastrophic. Further, a bug can be thought of as an intrusion or corruption in an otherwise flawless piece of program logic – bugs in the shape of bacteria *invade* an otherwise healthy body – whereas in reality a program bug might well be *inherent* in the problem-solving philosophy or conceptualisation.
>
> (Bloomfield 1989: 415)

Within this ideological frame, technical criteria continue to dominate organizational job design. In 1955 Davis *et al.* obtained information on criteria used by job designers and found that minimizing the time required to perform a task, obtaining the highest quality and minimizing skill requirements were seen as the most important factors. Even the more recent studies by Hedberg and Mumford (1976) and Taylor (1979) have revealed similar results. As a consequence, the relative benefits of human and machine control of work flows, work pace and job content are assessed in terms of increasingly quantitative criteria during design decision-making. Such an approach has long been recognized to bias decisions on allocation of functions in favour of automation. For instance, Jordan argues that designers tend to describe human functions in mathematical terms comparable to the terms used in describing mechanical functions. This classic example of a 'closed' circuit of meaning leads to a paradox whereby any time one can reduce a human function to a mathematical formula or algorithm, one can generally build a machine that can do it more effectively than a human: 'In other words, to the extent that man [sic] becomes comparable to a machine we do not really need him any more since he can be replaced by a machine' (Jordan 1963: 162).

Janice Klein (1991) observes that this labour displacement and objectification process has intensified as technological systems have increased in scale, complexity and interconnectedness. As the architectures of technological systems become more tightly coupled the scope for individual worker autonomy diminishes (see Chapter 3 and Corbett 1987).

USER PERCEPTIONS AND AFFECTIVE RESPONSES TO TECHNOLOGY

Users are generally seen as having an ambivalent view of technology. Thus, while tightly coupled technologies such as the car assembly line are often seen as de-skilling, even dehumanizing, users often welcome the introduction of new technology in general. What is clear from the research is that the study of user perceptions of technology needs to be informed by a consideration of (a) the wider socio-cultural ideology surrounding technological change, and (b) the ideological influences on the social construction of meaning at work, if this apparent paradox is to be resolved.

Early studies underplayed the wider socio-cultural context but evoke a strong sense of the anonymity of industrial technology felt by its shop-floor users. The classic study of *The Man on the Assembly Line* by Walker and Guest (1952) was one of the earliest social scientific examinations of the relationship between production technology and user perceptions and affective responses. The authors note that 'technological factors of automobile assembly work affect the worker ... directly through the immediate job, and indirectly by

modifying the basic organisational and social structure of the plant'
(Walker and Guest 1952: 20).

The study focuses on the impact of assembly-line technology on
the attitudes and affective well-being of car workers. The authors
conclude with the suggestion that

> the sense of becoming de-personalised, of becoming anonymous
> as against remaining oneself, is for those who feel it a psycho-
> logically more disturbing result of the work environment than
> either the boredom or the tension that arise from repetitive and
> machine-paced work.

(Walker and Guest 1952: 161)

In a similar vein to the work of Walker and Guest, Blauner's examination of the relationship between various production technologies
(craft, machine-minding, mass production and process production)
and workers' psychological well-being reveals a degree of (self-
reported) meaninglessness, social isolation and depersonalization
associated with work on the assembly line (Blauner 1964). He also
argues that such affective states are far less common among the
operators of highly automated process technology as such technology enables the development of team working, facilitates
operator control over the pace of work, and promotes the development of advanced technical skills and knowledge. Blauner maintains that this contributes to the operators' sense of belonging,
achievement and responsibility.

However, Zuboff (1988) and Gallie (1978) raise serious doubts
about the validity of this view in the light of more recent research on
the temporal constraints imposed on users by process technology.
Indeed, while there is general agreement among contemporary
researchers about the adverse psychological reactions associated
with assembly-line production work, a polarization of views has
emerged whenever researchers have turned their attention to more
technically sophisticated, computerized production technologies.
On the one hand, some researchers (e.g. Piore and Sabel 1984, Kern
and Schumann 1985) argue that advanced technologies herald a
break with the Taylorist/Fordist tradition, enabling progressive
upskilling and empowerment of technology users. For writers such
as Braverman (1974) and Cooley (1988), on the other hand, the
design and application of new technology threaten to displace and
de-skill users at an ever-increasing pace. Before examining this 'de-
skilling debate' in the next section, we first turn our attention to

some of the research findings on user perceptions and responses to such new technology in the workplace.

The 1984 Workplace Industrial Relations Survey (Daniel 1987) and subsequent analysis by Daniel and Hogarth (1990) show that perceptions of technology in UK industrial organizations align closely with the dominant ideology of the technology process. The focus of the survey was the use of new technology in relation to the industrial relations characteristics of the workplace. Worker reactions to three different types of workplace change were surveyed in over 2,000 public and private companies. These three types were:

1   *Organizational change* – referring to substantial changes in work organization or working practices not involving changes in technology.

2   *Conventional technical change* – the introduction of technology, not including new (i.e. microelectronic) technology.

3   *Advanced technical change* – the introduction of new technology.

The distinctions were made to enable the researchers to explore the extent to which change involving new technology differed from other forms of change which had long been common. Figure 5.1 illustrates the pattern revealed by management accounts of the reaction of manual workers affected by different change episodes of the three types. It is clear that the major contrast lay between reactions to technical change compared with organizational change. Daniel and Hogarth (1990) also discovered a similar pattern of results in the reactions of works managers, general managers, personnel managers and office workers.

Perhaps the most illuminating finding in Daniel and Hogarth's research relates to the implications of change for employment levels. It was discovered that technical change was so popular relative to organizational change that technical change resulting in the loss of jobs was supported substantially more strongly than organizational change leading to an increase in the numbers employed or stable numbers. Technical change also appears to act as a lubricant to ease the introduction of organizational changes that would have been less attractive to employees in other circumstances.

The authors carried out a number of case studies in order to investigate why technological change enjoyed so much more support than organizational change. The case studies confirmed that technical change was perceived by employees as intrinsically more attractive than organizational change and uncovered five main

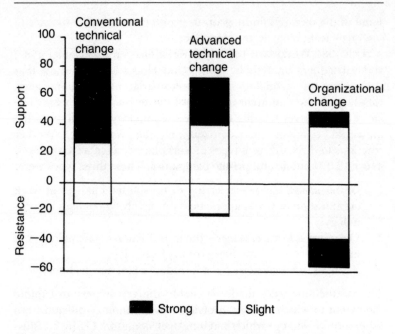

Figure 5.1 Reactions of workers to different forms of change
Source: Daniel and Hogarth 1990

reasons for this. First, new technology represented progress and advance; second, the benefits of new and improved machines were concrete, manifest and demonstrable; third, they represented competitive advantage – the modernity of its technology symbolized the standing of a manufacturing workshop in relation to others, both internally and externally; fourth, investment in new technology represented confidence in the future (and hence improved longer-term job prospects and security), and fifth, many of the features of new technology were familiar to and valued by workers in their domestic and leisure activities. Rolfe (1990) obtained similar results in her study of clerical office workers. These findings reveal that user perceptions of technology are influenced by ideological factors outside the immediate work environment.

But perceptions are also shaped by the experience of work itself, by pre-existing social relations and the management of meaning within the organization. In general terms, rules, regulations, and social, organizational and workgroup cultures (within the context of structural and ideological constraints) all serve to define how one perceives one's job and to provide possible sources of meaning in organizations. User perceptions of technology, while influenced by a socio-cultural ideological frame, will also be reflected in and reinforced by work identity – the personal meaning of work to the user. Research findings (e.g. Cook *et al.* 1982) suggest that work identity is closely interrelated with perceptions of user skills.

TECHNOLOGY AND THE NATURE OF USER SKILL

Since the publication of Braverman's *Labor and Monopoly Capital* in 1974, a great deal of research has been undertaken to investigate the relationship between new technology and skill. Unfortunately, the nature of the research findings serves to obfuscate more than illuminate the relative merits of the two sides of the de-skilling debate.

Consider the definition of 'skill' – the key dependent variable in research case studies. Most case studies cite the de-skilling hypothesis popularized by Braverman (1974) as their starting point and utilize similar definitions of 'skill'. But, as Attewell (1987) points out, Braverman developed his notion of de-skilling by contrasting two ideal types, a 'craft worker' and a 'detail worker'. The former requires a broad range of abstract theoretical knowledge via formal training which, in combination with manual dexterity, is used to plan and execute a variety of tasks under the craft workers' own direction. 'Detail work', on the other hand, can be learned quickly and requires no planning or theoretical knowledge. Also, it is routinized and closely supervised.

The ambiguity here is that this formulation of skill combines several dimensions to define skilled versus degraded work and these dimensions are themselves interrelated in a complex manner. For example, specialization does not necessarily imply a lack of skill (Littler 1982) and neither does routinization (Kusterer 1978). Furthermore, computerization tends to go hand in hand with abstract knowledge demands on the part of the user whether or not craft work or detail work is being carried out (Zuboff 1988).

Hence, the nature and type of skill may change with the application and use of new technology, but it is not possible to conclude that such a change is either de-skilling or enskilling without a clear point of reference. To assess this two-way process, what is required is a more precise understanding of what 'skill' or 'de-skilling' entails from both an objective and subjective perspective.

In terms of identity – the personal perception and meaning of work – skill has both an exchange (i.e. objective) value and a psychological (subjective) value to the individual in an organization. Organized social groups of individuals (e.g. trade unions and professional associations) can secure for themselves a higher level of wages if the definition of their skills is socially constructed in such a manner as to enhance their status, non-substitutability and perceived contribution to the achievement of organizational objectives. The collective meaning of skill relates to one's wage bargaining power and thus can be seen to possess a political quality (see Chapters 2 and 3). At the more subjective level, research findings suggest that identities do not differ in a random way between individuals, but are directly correlated with the level of skill involved (Cook *et al.* 1982). The more skilled and professional the work, the more likely it will be valued as a medium for self-expression, achievement, fulfilment and status by the job-holder (Kornhauser 1965, Fox 1980).

These two dimensions of skill – in economic exchange and in the formation of identity – are not always commensurable and this is another contributory factor to the confusions within the de-skilling debate. Research by Kusterer (1978), for instance, shows how car assembly workers (whose skill is objectively labelled as 'unskilled' by employers) perceive themselves as 'skilled'. Conversely, operators of manual machine tools often view computerized machine tools as degrading their skills even though their contribution to the labour process is formally recognized (and financially rewarded) as 'skilled' by their employers (Rubery 1978). Finally, conventional definitions of skilled work are biased in favour of white male skill such that, 'When a white man developed manual dexterity, it became a skilled trade, when a woman or black developed manual dexterity it was a natural characteristic and classified as unskilled' (Game and Pringle 1983: 7–8).

Hence, the 'answer' to the de-skilling question will depend, at least in part, on whether one concentrates on the subjective or the objective meaning of skill and the status accorded to the influence of ideology. Moreover, further complications arise from research evidence indicating that subjective skill assessment and evaluation are also influenced by the way in which the meaning of skill is socially constructed in the workplace and beyond. For example, empirical research following the 'social information processing approach' to work design (Salancik and Pfeffer 1978) reveals that job perceptions do not necessarily reflect objective job characteristics but are influenced by informational cues provided not only through the wider ideological enframement of technological thinking but by other people within the organization (see, for example, Griffin 1983, O'Reilly and Caldwell 1985, Griffin and Bateman 1986). Hence, in organizations where cosmetic 'employee empowerment', cultural change or similar techniques of attitudinal restructuring occur alongside technological change, workers' perceptions of their skill can be influenced to a significant degree by top management's attempt to manage the meaning of work (Silver 1987). In such organizations, machine tool operators, for example, can find themselves contributing to production brainstorming or quality circle sessions and be referred to as 'builders', 'owners' or 'task managers'. Such labels imply a degree of skill upgrading which may not be reflected in the nature of the work they carry out.

TECHNOLOGY, SKILL AND IDENTITY

What, then, is the impact of technological change on skill and identity? Zuboff (1988) argues that computerization of the production process restructures the nature of work. Computers are seen to abstract thought from action in so far as 'action-centred' skills are becoming replaced by 'intellective' skills. As a result, abstract theoretical knowledge (historically the domain of management) becomes of paramount importance to the user. The introduction of a computer as a mediator in the relationship between user and technology simultaneously entails a 'distancing' of that relationship.

> The construction of meaning from the electronic text that now represents the production process is likely to require more deliberate, controlled, aware, cognitive effort than earlier action-centred, context-dependent routines. The reason involves the nature of the inference process that links environmental cues with a meaningful response. When those cues are primarily concrete, the inference process can remain relatively implicit or tacit. When those cues are primarily abstract, meaning must be constructed through the explication process that assigns referents and analyses relationships.
>
> (Zuboff 1988: 191)

A key aspect of human skill is the ability to impose one's own definition and meaning on the tools, techniques and technologies being used (Corbett 1989). Research evidence (e.g. Singleton 1978) power-fully suggests that 'action-centred' craft skill is not amenable to an objectified predefinition of reality, however informal. Rather, a skilled user of technology utilizes an almost infinite repertoire of behaviour from a finite behaviour experience involving a continuous redefinition of reality. Such skill is essentially proactive in the sense that the human plans actions and carries out these actions – paying constant attention to their impact in terms of both feedback and feedforward, making the necessary adjustments to both plans and actions moment by moment.

However, 'action-centred' skill can also be applied in a reactive sense. For example, routine machine tool performance monitoring (machine minding) requires vigilance under strong temporal pressure and timely corrective action – usually hitting the 'panic button' – when errors arise during the machining process. But such skills are of a different order to proactive skills requiring preplanning, such as manual machining or complex parts assembly. Similarly, 'intellective skills' can be of a proactive (e.g. NC programming) or reactive (e.g. monitoring a sophisticated chemical processing system) kind.

The essential difference between reactive and proactive control over work process technology is neatly encapsulated in the following quotation from Karl Marx's *Capital*:

> A bee puts to shame many an architect in the construction of its cells. But what distinguishes the worst of architects from the best of bees is namely this: the architect raises in his imagination the structure he will ultimately erect in reality. At the beginning of every labour process we get a result that already existed in the consciousness of the labourer at its commencement.
>
> (Marx 1954: 174)

In broad terms then, proactive control over technology involves the actor imposing her own meaning on her interactions, whereas reactive control involves the following of instructions imposed by others in the design and programming of that technology, whether or not the skills in use are of an 'action-centred' or 'intellective' kind (see Figure 5.2). In a highly automated system where the operator serves the function of 'closing the control loop', human actions tend to be directed and prompted by cues and instructions embedded in software and hardware, i.e. the user *reacts* to the technology rather than acting upon it. Returning to Marx's analogy, the architect acts upon the world, whereas the bee responds according to the 'programming' embedded in its genetic makeup.

When this framework is mapped onto the findings within the 'de-skilling' debate, a degree of consistency becomes manifest. While it is clear that 'action-centred' skills are being increasingly displaced by 'intellective' skills as computer technology develops, the subordination of the user to technology is only complete when a corresponding shift from proactive to reactive control skills occurs.

It follows from this two-dimensional view of de-skilling that, in general terms, the process of technological change 'must be seen as simultaneously deskilling and reskilling the labour force. Rather than a simple one-way process that Braverman describes, we must recognise this more complicated, two-way process' (Edwards 1979: 121). Furthermore, while new technology is associated with a skill shift towards a more abstract application of knowledge and skill, this does not necessarily lead to either a 'reprofessionalisation' of the shop-floor worker as Kern and Schumann (1985) maintain, or to de-skilling. One must ascertain whether the worker has a proactive or reactive control position within the context of the technical control

LOCUS OF SKILL

	Action-centred	Intellective
Proactive	e.g. Craft work	e.g. Machine or systems programming/ planning
Reactive	e.g. Machine minding	e.g. Systems monitoring

LOCUS OF CONTROL

Figure 5.2 Profiles of skill in the use of technology

system. Research evidence suggests that, although some production jobs with new technology require a degree of proactive skill (e.g. operator programmable CNC machine tools), a significant majority of jobs require primarily reactive skills (Zimbalist 1979, Thompson 1989).

Indeed, the psychological reactions of users in jobs requiring reactive 'intellective' skills are similar to those expressed by assembly-line workers, utilizing reactive 'action-centred' skills, in earlier studies (Walker and Guest 1952, Blauner 1964) in all but one important respect. In effect, the psychological and physical distance brought about by the computer mediation of the production process renders the latter relatively opaque to the user. Consider the following quote from a pulp mill worker:

> Our operators did their job by feeling a pipe – 'Is it hot?' We can't tell them it's 150 degrees. They have to believe it. With computerisation I am further away from my job than I have ever been before. I used to listen to the sounds the boiler makes and know how it was running. I could look back at the fire in the furnace and tell by its colour how it was burning. I knew what kinds of adjustments were needed by the shades of colour I saw. A lot of men also said that there were smells that told you different things about how it was running. Now I only have numbers to go by. I am scared of that boiler, and I feel that I should be closer to it in order to control it.
>
> (Quoted in Zuboff 1988: 63)

In consequence, computerization of organizational work processes has two important consequences. First, the 'psychological distancing' identified by Zuboff engenders a sense of dependency on technology as most, if not all, sensual information is replaced by computer-generated data – algorithms being the objectification of workers' 'action-centred' know-how. As a result, the possibility exists that workers will develop a kind of 'learned helplessness' and cede all effective control to automation.

Yet, the process of identity formation suggests that workers may not passively accept the functional definition of the 'user' embedded within the 'closed world' ideology of the engineering manager and systems designer. Indeed, being defined in terms of a relatively powerless relationship to technology or to the 'system' may encourage workers to develop identities based on opposition to that system. Conversely, just as users may hold an ambivalent view of technology, the realities of the use-process may lead designers to hold an ambivalent view of the user. It is precisely because users do not always function 'correctly' or 'efficiently' within the designers' frame of reference that designers often call upon systems logic to argue that automation should displace the worker altogether. Yet, in the interim, the designer remains dependent on the (often tacit) knowledge and skills that inform much of this 'deviant' user behaviour to ensure that the system designed functions effectively. Indeed, should users conform rigidly to a 'closed world' definition of their role – that is, by 'working to rule' – systems performance would degrade rapidly. Such dependence on user know-how is all the more poignant, of course, given that the objectification of this knowledge is a fundamental aspect of technological innovation.

A second consequence of computerization comes to light from the research evidence pointing to the low levels of job satisfaction and motivation associated with jobs requiring a reactive 'intellective' skill profile. Such job-holders have what may be termed 'negative responsibility', i.e., they have job responsibility, but possess little actual or perceived work autonomy (see our discussion of job design in Chapter 3). Our earlier discussion on the objectification of time (Chapter 4) shows how the ideological frame of technological systems even constrains users' ability to define work time.

SUMMARY: TECHNOLOGY AND CIRCUITS OF MEANING

In this chapter we have argued that while the perceptions of management, designers and users differ in important respects, they are often equally enframed by ideologies of scientific and techno-logical rationality. Such ideologies constitute not only a resource which powerful groups can call upon to legitimate the work innovations they initiate, but also help to shape the perceptions of skill and work identity of all groups in the technology process, designers and engineers included. Thus, to the extent that such ideologies and meanings are seen to have contradictory or dysfunc-tional consequences for the development and use of technology – in worker reactions to the status of 'user', for instance – such conse-quences are not amenable to functional interventions alone (namely, programmes of 'user involvement'), but bear on more intractable issues to do with the identities and perceptions of those involved in the technology process.

Part III

Design

Give me somewhere to stand and I will move the earth
Archimedes, speaking of the lever

When the electronic circuitry is etched into the silicon surface of a microchip, the resulting pattern of connections and gates transforms the chip from a worthless piece of mineral into a potentially powerful information processing technology. Technological knowledge, economic decisions and practical insight have been condensed into a specific design.

This kind of transformational power simply cannot be understood in terms of political or ideological forces alone, though it is often deployed to support such forces. To speak of an organization's use of technology simply in terms of the 'intentions' or 'values' of one group or another is to understate the extent to which organizations and technologies transform each other over time, often in unintended ways.

One way of understanding this transformation is through the notion of design. Although the design of technology is commonly analysed under the headings provided by specialist technical activities (product design, systems development, process implementation), in sketching out the analytical framework for this section we aim to avoid presenting it as a specialist function which stands outside the organizational mainstream. Rather, we see it as an activity that links organizations, and particular groups within them, into the technology process. Moreover, although the literature on design and innovation tends to conflate the normative with the descriptive, confusing 'should be' with 'is', in reviewing that literature we will not lose sight of the political and ideological forces that were described in earlier sections. Indeed, the kind of overriding concern with efficiency and effectiveness that encourages the conflation of analysis and prescription can also be viewed, in the terms of our earlier discussion, as the political and ideological frame within which new technologies and new ideas are evaluated. Within this frame, the relative efficiencies and utilities created by new technologies not only become their own validation, but also determine the perceived success or failure of the groups and organizations that promote them.

Analogous with the signals transmitted through an electronic circuit, design can be seen as involving flows of knowledge, information and artefacts between the processes of invention, exchange and use. As design activity shifts from one level of technology to another – from generic knowledge to specific implementations – the content of those flows gradually changes until a final assembly of artefacts, techniques and knowledge is defined.

The first task of analysing technological design is to understand the nature of the signals flowing through the circuit, and the ways in which they are generated and received. However, if we restricted our analysis simply to these outward signals of design activity – the decisions based on specific pieces of information, for instance, or the layout of a particular technical system – we would be neglecting the underlying structures that shape the flows of knowledge and information.

Our second task, therefore, is to analyse the structure of the circuit itself. Like the paths and gates of the silicon chip, the knowledges held by various groups within the organization – and the relationships between those knowledges – shape the flows of artefacts and information both inside and outside the organization. They constitute an underlying *design circuit*, out of which specific technological and organizational implementations emerge.

Chapter 6

Design, knowledge and innovation

INTRODUCTION

This chapter will consider design activity in terms of the generation, transmission and assembly of knowledge, information and artefacts. Analysing design activity in these terms is not to view it as a simple chronological sequence where one phase follows another. This combination of new and existing forms of knowledge, which is the distinctive feature of design, may take place in a variety of social contexts and is hardly ever a neatly sequential process. It is shaped, however, by the specific context of invention, exchange and use. Putting it simply, the design of a car, say, might be influenced by a variety of contingencies:

1 The state-of-the-art knowledge on bodyshell materials.
2 The projected costs of labour and components, and market research on the particular customer segment at which the car is aimed.
3 The projected performance of the car and the kinds of road surface on which it will be driven.

However, the empirical 'facts' on which the design is based do not present themselves as an unproblematic external reality. Even scientific 'facts' are dependent upon social consensus and complex processes of interpretation (Collins and Pinch 1982). Similarly, the knowledge and information which inform design activity are the products not of disinterested observation but of purposeful activity. The remainder of this chapter is devoted to understanding how the activity of different groups creates and sustains the knowledge and models of external reality that underpin particular design projects.

GENERATION

Design requires stimulation and focus. Sometimes this may come from the scientific field; the development of solid-state electronics owes something to advances in solid-state physics, for instance (Braun and Macdonald 1978). Sometimes it comes from the prompting of economic forces as expressed by measurements of productivity or costs. However, even economists admit that economic forces on their own do not provide a strong direction or stimulus for technological change:

> the ultimate incentives are economic in nature; but economic incentives to reduce cost always exist in business operations, and precisely because such incentives are so diffuse and general they do not explain very much in terms of the particular sequence and timing of innovative activity.
>
> (Rosenberg 1976: 110)

Very often, it seems, the most direct stimulus to technological design comes from problem-solving activities centred on the use of existing technology. Thus the development of solid-state electronics owed as much to the possibilities thrown up by wireless telegraphy and valve electronics as it did to pure physics (Braun and Macdonald 1978). Different patterns of use throw up different problems and puzzles to be solved, leading technological development in one direction and excluding others. Bronowski notes of the Incas in Peru, for example, that although they developed the use of rollers for moving large stones, this never led them to the invention of the wheel. As he says, 'what is radical about the wheel is the fixed axle' (Bronowski 1973: 101). The Incas' use of rollers simply never led them into attempting to solve this particular technological problem.

On the other hand, those technological bottlenecks – what Hughes (1987) terms 'reverse salients' – that do become visible can play an important role in guiding the attention and resources allocated to design activity.

In the computing field, for instance, the three major phases of development have each been dominated by a particular configuration of technical and organizational problems:

1   Late 1940s to mid-1960s:
    Hardware constraints phase. Dominant problems were the costs of hardware and its limited capacity and reliability.
2   Mid-1960s to beginning of 1980s:

Software constraints phase. Focus on the 'software crisis' – the low productivity of systems developers – and the difficulties of delivering reliable systems on time and within budget.

3   Beginning of 1980s to present:
     User relations constraints phase. Systems quality problems arising from inadequate perception of user demands and inadequate servicing of their needs.

(Friedman and Cornford 1989)

As problems emerge out of existing technology, the problem-solving focus encourages an incremental and cumulative pattern of technological change. This creeping form of technological determinism is limited, however, by the influence of social and economic forces on perceptions of the problem focus.

Social expectations about future advances in technology can play an important role in shaping the problem focus (MacKenzie 1990). A perceived dominant problem sucks in resources and commitments to such an extent that social demands and expectations may become a self-fulfilling prophecy. This seems to have been the case, for instance, with the early development of the integrated circuit in the 1950s. The demands of the US military for more and more miniaturization were fuelled by events such as the Soviet Union's launching of the Sputnik in 1957. They certainly were not scientific demands. According to one scientist: 'All that you had to do was to wave the Russian threat and you could get money' (quoted in Braun and Macdonald 1978: 109).

The problem-solving focus is not simply a function of general perceptions, however. There are significant checks and balances on the power of social expectations. These are graphically demonstrated when radical shifts in the economic or scientific environments cause the dominant problem to switch unpredictably from one element to another. This is the moral provided by the contrasting stories of the designers of Concorde and the Boeing 747, outlined below.

Case example: Dominant design problems with Concorde and the Boeing 747

In the late 1950s and early 1960s, when the Anglo-French Concorde was being designed, the dominant problems of air passenger transport were speed and distance. Compare this with the situation in the

1970s when Concorde was actually being introduced. The 1973–4 oil crisis suddenly made fuel efficiency and hence sheer carrying capacity the critical design parameters for aircraft manufacturers.

The aircraft that most effectively met the new design parameters was not Concorde, but the Boeing 747 'Jumbo Jet'. However, with an aeroplane design of this size and weight – three-quarters of a million pounds – a new dominant design problem emerged. As the 747's designers discovered, the problem was not whether the plane would fly, but rather how to land it at speed. In the new era of mass air transportation, it transpired, the critical design bottleneck was not to do with speed and distance but rather the problem of turning a mass transit aeroplane into a high-speed, heavy-wheeled land vehicle. The success of the 747 ultimately hinged on the company's ability to develop elaborate, 20-wheeled, landing gear technology which resolved the landing problem.

Source: Gardiner (1986)

TRANSMISSION

The transmission of knowledge and information rests on the manipulation of symbols and other forms of representation. The latter may actually take a physical form. Artefacts, for instance, may help to transmit knowledge heuristically. In the post-war arena of missile technology, both the USA and USSR used the German V2 rocket as a model for their own research and development. At the end of the war, the Americans had taken possession of a number of V2s (not to mention a goodly number of V2 designers), while the Russians had to resort to painstaking reconstructions of the original design (MacKenzie and Wacjman 1985). More generally, concrete examples of the 'state of the art' provide an important starting point for the product design process. In a perverse replay of the assembly of knowledge and artefacts that goes into a product design, designers will often begin the development process by literally 'deconstructing' the 'state of the art' product they aspire to emulate or outdo.

Information is also transmitted through less concrete symbols which are harder to deconstruct. Forms of language, measurement and control can have a significant effect on the transmission process. A significant factor in the UK, for example, is the role played by financial accounting measures. This reflects the power exercised by accountants within management structures generally, and their superior competence – relative to personnel managers, say – in measuring and controlling business activities. The language of accountancy turns technological design into a matter of return on investment, cash flows and productivity gains (see Wilson 1992). Proposals for technological design have to be translated into this language, albeit reluctantly, by specialist technical groups. Although such exercises are often ritualistic or highly political in nature – Freeman (1974) likens the financial evaluation of technology to 'tribal war dances' – by establishing a language in which design proposals have to be transmitted they condition the character of the design activity itself.

But, while the language of accountancy is an effective means of transmitting the more abstract aspects of design, as activity shifts towards the use-process itself, other methods and symbols come into play. A notable example is provided by the use of 'methodologies' in systems development and operations research.

The problem here is finding a way of translating the complex social reality of the use-process into an objective systems specification. With the earliest applications of computing, this proved to be no great problem. The first computer-based systems were actually direct replacements for manual ones, and consequently could be specified entirely through existing systems descriptions (Friedman and Cornford 1989).

However, as computer systems were applied to more complex

Table 6.1 A comparison of the 'real' and conventional systems development cycle

The 'real' systems development cycle	The conventional systems cycle
1 Wild enthusiasm	1 Feasibility study
2 Disillusionment	2 Requirements analysis
3 Total confusion	3 Systems analysis
4 Search for the guilty	4 Specification
5 Punishment of the innocent	5 Design and development
6 Promotion of non-participants	6 Implementation

Source: Taggart and Silbey 1986

organizational tasks, the focus of methodologies shifted. Narrow
concerns with the design of hardware gave way to 'softer' methods
aimed at eliciting the subjective knowledge and perceptions of the
user. These seek to provide a 'rich' picture of reality based on more
active participation by users and a consensus on agreed changes.

Even so, there remain significant limitations on the extent to
which such methodologies are able to represent and communicate
the complexity and uncertainty of the use-process for technology.
The gap between such abstract, objective models and the political
and subjective reality of implementation is still big enough to
prompt hollow laughter from many of those involved, as Table 6.1
demonstrates.

ASSEMBLY

The assembly of different forms of knowledge, technique and
physical material into a specific implementation is a costly and time-
consuming exercise. This is demonstrated most visibly in areas of
high product complexity such as the aerospace and car industries
where the costs of product design have prompted radical restructuring, mergers and collaboration. Manufacturers have been forced
to develop 'robust designs' whose basic features can be 'stretched'
over time to encompass changing customer needs. Product designs
such as the Ford Cortina, the Boeing 747 and the Canon photo-
copier fall into this category (Gardiner and Rothwell 1989). Others,
notably the Comet and Concorde airliners, certainly do not.

Similar motives prompt organizations to develop designs based
on the reassembly of existing components and knowledges. The
Black and Decker paintstripping heatgun, for example, is derived
from the company's standard electric drill. Black and Decker used
the motor, fan, case and switch from proven drill components, and
replaced the transmission and chuck assemblies with a new heater
element and nozzle (Gardiner and Rothwell 1989).

By exploiting existing knowledges and technical components, the
gradual reassembly of technology allows organizations to economize
on both the cost and the uncertainty of technological design. At the
same time, however, it involves accepting important constraints on
the final shape of the technology. This is particularly evident in the
addition of new technological components to existing infrastructural
systems.

Current developments in electronic banking in the UK provide a

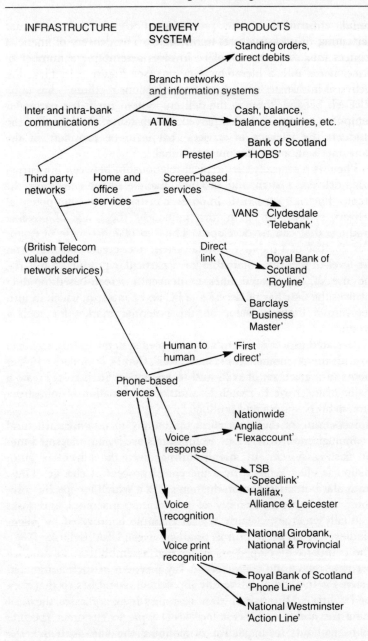

INFRASTRUCTURE

DELIVERY SYSTEM

PRODUCTS

Standing orders, direct debits

Branch networks and information systems

Inter and intra-bank communications

ATMs

Cash, balance, balance enquiries, etc.

Bank of Scotland 'HOBS'

Prestel

Third party networks

Home and office services

Screen-based services

VANS   Clydesdale 'Telebank'

(British Telecom value added network services)

Direct link

Royal Bank of Scotland 'Royline'

Barclays 'Business Master'

Human to human

'First direct'

Phone-based services

Voice response

Nationwide Anglia 'Flexaccount'

TSB 'Speedlink'

Voice recognition

Halifax, Alliance & Leicester

National Girobank, National & Provincial

Voice print recognition

Royal Bank of Scotland 'Phone Line'

National Westminster 'Action Line'

Figure 6.1 Infrastructural technologies: the assembly of home and office banking systems in the UK

useful illustration of the problems of choice and constraint pertaining to infrastructural technologies. The delivery of financial services into the home or office involves assembling a number of components into a hierarchical design (see Figure 6.1). First, the technical infrastructure – the communications medium – has to be selected. Second, there is the delivery system itself: the particular computer system to be employed, for instance. And finally, the product: the package of services that is to be provided on the consumer's phone or computer terminal.

There is a logically hierarchical relationship between infrastructure, delivery system and product. Each element constrains the other. The infrastructure imposes constraints on the potential delivery system, which in turn influences the range of possible products that can be developed. Thus, to take a couple of examples, the use of a public viewdata system may create constraints on the level of security obtainable for a particular product. Similarly, the use of, say, leased lines to deliver a screen-based product involves the user having access to a PC and a modem, which in turn becomes a limiting factor on the potential market for such a product.

A related aspect of assembling complex systems in this way is to do with the phenomenon of 'sunk costs', that is, cumulative investments in a given set of skills and technologies. Such costs create a major disincentive to switch to another combination of infrastructure, delivery system and product.

Something of the same effect may apply to non-infrastructural technologies too, where new technologies are being integrated into an existing system. In manufacturing, even the advent of automation is often typified by a piecemeal process of change. Thus, particular bottlenecks in production spark a search for specific solutions, and a new assembly of techniques, machines and tasks gradually emerges. A fairly typical example is provided by Allen-Bradley, an industrial controls producer in the USA (Bylinsky 1986). The company developed an automated assembly line to produce electrical relays and contactors. A key problem in such automated systems is the ability to identify individual assemblies so that they can be directed through different manufacturing processes, the ideal being the ability to process individual units to customer specifications without slowing down or stopping the line. Although the company had made a massive investment in process machinery, the solution to this key problem came from the simple idea of transferring the kind of bar-code labels that are extensively used in the retail trade to an industrial application. Assemblies could thus be quickly and individually identified by their specific bar-code. Moreover, in order to make bar-coding a possibility, Allen-Bradley engineers came up with the idea of using high-speed laser-printers to formulate and print bar-codes on the factory-floor.

This piecemeal pattern of change applies with equal effect to the design of work organization associated with new technology. Having picked off and automated the most routine, robot-like jobs, the existing patterns of work organization are often left, relatively unchanged, to simply flow around the newly established 'island of automation'. The dominant design principle here seems to be one of 'least pain, greatest gain': avoiding any major rethinking of the basic system design, while securing the immediate productivity benefits of new technology.

At the same time, however, there is evidence that such pragmatism may create a false economy of thought and effort. The failure to redesign technology and organization together – which is described in more detail in Chapter 3 – has been widely criticized for neglecting the social factors that lie outside the immediate hardware domain. This is not simply a question of the pursuit of the quick 'technical fix', though Rothwell (1984) notes how, in the UK at least, the financial benefits of rapid implementation often override long-term planning for technology. Rather, knowledge about the social aspects of technological design simply does not gain access to the design process (see Chapter 5).

Heller describes how the systems experts who designed the UK's Driver and Vehicle Licensing Centre (DVLC) in Swansea pursued a Tayloristic approach to work organization: 'each operator had a very narrow, unskilled job to carry out: opening post; examining contents; editing; coding; applying batch numbers; microfilming; key-punching etc' (Heller 1989: 188). The result was a highly inefficient system. The original design estimates indicated a cost of £146 million and 4,000 employees, but the reality ballooned to £350 million and 7,900 staff.

Apart from the sheer drudgery of the work that they produce, such designs are subject to the 'socio-technical systems' critique (Hedberg and Mumford 1976) that they attempt to optimize the technical system while neglecting the social system. Although this approach can itself be criticized for neglecting the 'political system' of the organization (as we discussed in Chapter 3), it does offer a more balanced model for bringing together social and technical forms of knowledge in the design process.

In the DVLC case, and many others, it was managerial dissatisfaction with the performance of the Taylorist system that prompted the incorporation of 'human factors' into design. Telephone and correspondence work were integrated, and workers organized into flexible work-teams of eighteen members. Significant improvements in worker satisfaction and operational performance were achieved and sustained (Heller 1989).

Ironically, however, even the apparently optimistic outcomes of the DVLC case may inspire a rather gloomy prognosis about technological design in general. For this case only serves to underline the extent to which Tayloristic assumptions and knowledge remain preeminent with designers for reasons other than their inherent efficiency or rationality. While many other cases before the DVLC design had clearly demonstrated the counter-productive aspects of a Taylorist approach to work organization, this did not prevent the DVLC designers from embracing Taylorist models wholeheartedly in the original design. Nor did either precedent or even current experience prevent them, as Heller (1989) notes, from persisting in the application of Taylorist precepts – segregating all the telephones into one specialized area, for example – even as the socio-technical consultants were trying to break down such barriers in other areas.

This kind of recidivism could be interpreted simply in terms of the 'closed system' thinking of process engineers and designers. Yet, the persistence of particular forms of knowledge and rationality in the face of overwhelming evidence to the contrary is too general a feature of technological design to be adequately explained in these terms alone. It is demonstrated in its most extreme form by the case example of the M-16 rifle.

Case example: The M-16 rifle in Vietnam

At the beginning of the Vietnam War in the mid-1960s, US combat troops were supplied with the M-14 rifle. This was essentially an automatic-firing version of the previous standard issue M-1 rifle. The development of the automatic-firing version had been encouraged by research into the behaviour of US combat units in the Second World War. This had revealed that four out of five soldiers equipped with the M-1 – a design that required a separate squeeze of the trigger for each round fired – did not actually fire their weapon during a battle. This was contrasted with the behaviour of troops with automatic rifles who were much more likely to use their weapons. The contrast was explained in terms of the psychological sense of control – the ability to dominate an area by 'hosing it down' with bullets – that the automatic weapon provided.

Although the M-14 was automatic, it had inherited its basic design from the M-1 and consequently was cumbersome to use, being virtually uncontrollable in automatic firing mode. The dissatisfactions of the users of this technology – the troops on the ground in Vietnam – were eventually communicated back to the US Dept of Defense. The order went out for a new weapon to be procured for combat use in Vietnam. The original choice of the combat troops was the Armalite AR-15, an automatic rifle which was both highly lethal and highly serviceable in terms of reliability and portability.

Control over weapons procurement lay with the Army Ordnance Corps – 'an informal congeries of weapons laboratories, private contractors, and the Army Materiel Command' (Fallows 1985: 241). This highly conservative body had developed its values and technical standards on the premise that marksmanship – the ability to hit a target 400 to 500 yards away – was a critical feature of weapon performance. In reality, most fire fights in combat took place at a range of less than 50 yards. However, given its concern with marksmanship, which perhaps masked a more general NIH (not invented here) syndrome, the Ordnance Corps took a dim view of the AR-15. In particular, the Corps found the rifle deficient in relation to a key technical standard: given its current design and ammunition, and over a temperature range that went from an Arctic −65°F to a tropical +125°F, the AR-15 just failed to attain a bullet velocity of 3,250 feet per second.

Despite the practical advantages of the AR-15, and the irrelevance of the Arctic temperature test to its performance in the jungles of Vietnam, the Ordnance Corps' dominance over the design process ensured that its standards and values prevailed. Consequently, the AR-15 was heavily modified to satisfy Ordnance Corps standards, and various new and largely unnecessary features were bolted onto it. Most importantly, in order to reach the bullet velocity standard, the rifle's ammunition was changed.

The effect of these modifications was drastic. Unlike the highly reliable AR-15, the new rifle was prone to fouling and jamming, which often had fatal results for combat troops in the field. Many soldiers wrote back to their parents to complain grievously about such problems. A typical letter read: 'I was walking point a few weeks back and that piece of you know what jammed three times in a row on me. I'm lucky I wasn't doing anything but reconning by fire or I wouldn't be writing this letter now.'

When jammed, the new rifle could not even be used as a club because it had been designed with a plastic stock for lightness. It was reported that after battles the Viet Cong would strip the American dead of everything but their rifles, which they considered worthless.

Source: Fallows (1985)

SUMMARY

The explanation for the technological outcomes that we find in cases like the DVLC or the M-16 rifle seems to lie less with the misguided nature of designers and engineers than with the particular position that they hold within the organization. There seems to be an underlying structure to design activity which is to do with the distribution of technological knowledges between different organizational groups, and the relationships between those knowledges and groups.

This underlying structure – which we have termed the 'design circuit' – helps to explain the way in which the US Ordnance Corps, like many expert groups before it, was so comfortably insulated against even the most desperate pleas of those who had to use or live with a particular technological design.

Chapter 7

The design circuit

INTRODUCTION

The knowledges possessed by different groups inside and outside the organization play a key role in design activity. Earlier, we described how design activity is to a large extent a matter of assembling together the knowledges developed by different groups. Specifically, it entails bringing together new knowledge with knowledge of an existing technology process.

But, as earlier sections have indicated, the distribution of knowledge and skill between different groups is influenced by political and ideological factors. Each group has an in-built incentive to defend and enhance its control over the knowledge relevant to design. Losing control may mean that the group's strategic power is undermined by any subsequent technological change.

Moreover, knowledge is not only 'sticky' (von Hippel 1990) in a political sense. Beyond the codified generic knowledges that inform design, there may be a wide array of subjective and tacit forms which are not readily communicable, but which influence design through the subjective involvement and control of different expert groups.

What these political and subjective factors create is a silent shaping of design activity through an underlying and persistent structure of knowledges and relationships. We have termed this the 'design circuit' to contrast it with the more visible signals of knowledge and information that are thrown out during a specific design project.

As a first step to understanding this circuit, we can usefully compare it with the conventional models of product and process innovation. The latter (see Table 7.1) tend to focus on the generation

Table 7.1 Models of product and process innovation

Product innovation

Idea generation	Conceptual definition	Product development	Pilot production	Marketing
Central R&D	Divisional R&D	Divisional engineering	Divisional production	Divisional marketing

(Shrivastava and Souder 1987)

Process innovation

Basic research → Development → Engineering → Use
            ADOPTION   IMPLEMENTATION

of new knowledge and its gradual translation into technological hardware.

According to such models, different specialist knowledges are gradually assembled together in a linear and sequential manner reminiscent of the Ford assembly line. Unfortunately, the conceptual assumptions of linearity and orderliness have begun to seem quite as rigid and inflexible on the theoretical plane as Fordism is on the shop-floor.

The theoretical objections are many and various. First, by emphasizing the generation of new knowledge through in-house research, they neglect the importance of the knowledge generated by users of the technology, many of whom may be outside the producer organization. One survey of the scientific instruments industry found that 80 per cent of the major technological improvements were originated by the users and not the suppliers of the instruments (von Hippel 1976).

Second, the linear model presents innovation as a neat and tidy activity when, in fact, it is characterized by multiple iterations and wrong turnings (Rothwell 1986). Nor is innovation completely divorced from existing technology. A good deal of the technological improvement in products and processes is actually the product of incremental rather than radical innovation.

In sum, while the model usefully highlights the various forms of knowledge and information required at different levels of technology, it tends to neglect the importance of feedback from existing users of the technology, be they user companies or shop-floor workers. Most importantly, its central premise – that specialist knowledges can be brought together in a logical and sequential manner – seems unrealistic.

Indeed, where Life has imitated Art, and organizations have managed design activity in this linear way, the results have often been far from satisfactory. In design activities as diverse as new product development (Shrivastava and Souder 1987) and the implementation of new technology (Leonard-Barton 1988), the linear transfer approach imposes a sort of order on design activities but only at the cost of political rigidities and poor interfunctional communications.

Where design activity is focused on complex and dynamic production processes and markets, the sequential approach has often been abandoned in favour of more 'holistic' approaches which recognize that design is 'a constellation of processes involving many players' (Clark and Staunton 1989: 196).

ORGANIZATION STRUCTURE AND KNOWLEDGE

One of the earliest studies to recognize the importance of viewing design and innovation in a broader organizational context was the work of Burns and Stalker on *The Management of Innovation* (1961). They contrasted the implications of 'mechanistic' and 'organic' structures for the conduct of technological innovation. The latter provided a supportive context for innovation by allowing the free flow of knowledge and information to override existing job designs and procedures, with no attempt to centralize knowledge at the top.

At one level, the Burns and Stalker analysis can be read as a classic piece of contingency theory, where the organic structure provides a more adaptive vehicle for organizations facing a dynamic environment. But, reframing the analysis in terms of the contribution of different knowledges to design, a different picture emerges. The organizations identified as organic saw themselves operating in a rapidly changing technological environment. Their central concern was the generation of new knowledge – existing structures could contribute little to this. The mechanistic organizations operating in stable conditions simply reversed this emphasis:

> the response to change was usually to create a new group, or to reconstitute the existing structure, or to expand an existing group which would be largely responsible for meeting the new situation, and so 'not disrupt the existing organization'.
>
> (Burns and Stalker 1961: 8)

It is easy to see this as simply an indictment of bureaucratic inertia, suppressing change and innovation. However, at some level prevailing structures may also communicate and embody the knowledges that shape design activity. Where organizations are engaged in a highly complex but stable set of tasks, it may be more important to transmit accumulated knowledge of existing products and processes than to open the organization up to generic knowledge of the changing technological environment.

Such accumulated organizational knowledge, both tacit and codified, is gradually acquired over time and widely distributed between different groups. It may become politically, physically (factory layouts, machine systems) and cognitively institutionalized within the organization. Forms of standardization and implicit ways of working may serve to communicate it. On occasion, even the hardware and products emanating from a design circuit may reflect such knowledge as a unique organizational imprint.

As an illustration, the following passage from *The Soul of a New Machine* (Kidder 1981) describes the reaction of Tom West, a computer designer with Data General, to a new 'state of the art' minicomputer produced by the rival DEC company. Having 'reverse engineered' the new VAX 11/780 machine – that is, taken it to bits – Tom West's reaction was critical:

> Looking into the VAX, West had imagined he saw a diagram of DEC's corporate organization. He did not like, for instance, the system by which various parts of the machine communicated with each other; for his taste there was too much protocol involved. He decided that VAX embodied flaws in DEC's corporate organization. That amazingly successful but cautious, bureaucratic style.
>
> (Kidder 1981: 36)

Once ingrained in a design circuit, such 'styles' can have important implications for a company's strategic development. For example, when Whipp and Clark (1986) examined the design process at the Rover car company, they found the organizational equivalent of Tom West's discovery. Going back into Rover's history, they saw what they call a 'structural repertoire' being institutionalized and embedded as far back as the mid-1930s. This 'repertoire' – the term denoting a limited range of responses to different situations – was the product of a mid-1930s period of organizational learning and economic crisis. It had been developed to connect an up-market niche for Rover models to a team management approach to product design and a craft-based use of technology. Although Rover were exposed to the same kinds of technological environment as the other producers, their structural repertoire strongly conditioned their use of new techniques. For instance, when the assembly line was introduced at Rover in 1946, it had little to do with management control or 'speed up', but was actually 'very slow ... a simple linking mechanism' (Whipp and Clark 1986: 70).

In the 1970s the new corporate group owning Rover attempted to transform their repertoire to one based on a new model aimed at a much larger European market and produced from a more Fordist work organization. The attempt proved disastrous: a car aimed at the European luxury car segment ended up with quality levels so poor that in one spot check only 6 per cent of sample output was found to be fit for immediate sale.

The Rover example demonstrates the overriding importance of the repertoire or design circuit in determining an organization's adaptation to a changing technological environment. Whatever the merits of a particular set of technological possibilities, their application rests upon an organization's ability to realize them through its design circuit. This may require a considerable degree of organizational learning, as well as the 'unlearning' of obsolete concepts or practices embedded within the circuit. As Whipp and Clark put it: 'the grasping of new opportunities means that the existing body of corporate knowledge has to be "creatively destroyed" and constructively replaced' (Whipp and Clark 1986: 213).

LINKING MECHANISMS

One important implication of the above examples is that the conditions under which knowledge is acquired have consequences for the way in which it can be deployed. Knowledge acquired through organizational learning may be transmitted and deployed through the institutional form of specialized functions and structures. Conversely, the acquisition of new forms of knowledge involves fluidity of structures and openness of both internal and external communications.

Hence, in attempting to develop design circuits based on a variety of new and existing knowledges, organizations are faced with a series of problems to which there may be no one structural solution. One typical organizational response is the recourse to 'ad hocery' in the shape of more or less ephemeral 'linking mechanisms'. These can be seen as a means not only of linking existing specialist functions, but also of connecting old organizational knowledge with new knowledge of the changing technological and market environment.

One such device is the development of the project team as a unifying focus for design activity. Although such teams may vary in the range of groups represented and the relative degree of commitment, the general aim is to provide a social context both for energizing and co-ordinating design activities, and for the pooling or trading of relevant knowledges. Depending upon the degree of autonomy it is given from the corporate structure, the project team may be able to develop a powerful group dynamic which thrives on the uncertainties and challenges posed by technological design. The following passage describes the powerful dynamics that infused Tom West's 'Eagle project' team at DEC:

> There was it appeared a mysterious rite of initiation through which, in one way or another, almost every member of the team passed. The term that the old hands used for this practice ... was 'signing up'. By signing up you agreed to do whatever was necessary for success. You agreed to forsake, if necessary, family, hobbies and friends – if you had any left (and you might not if you had signed up too many times before). From a manager's point of view, the virtues of the ritual were manifold. Labour was no longer coerced. Labour volunteered.
>
> (Kidder 1981: 63)

However, as the Eagle example demonstrates, project teams are often most successful where the grouping is of like-minded technical specialists. In the UK and USA, there have been relatively few attempts to develop completely interdisciplinary teams, and where such teams have emerged their members have typically been on short-term loan from specialist departments (Womack et al. 1990). In contrast, the Japanese have been able to exploit the more generalist background of their technical and managerial cadres to great effect in developing genuinely interdisciplinary teams. The Japanese practice of rotating managers between functions may create expertise shortfalls in certain high technology areas such as advanced computing technology, but it does reduce the constraints of specialist expertise and allows a much more holistic and integrated approach to multidisciplinary projects. This allows conventional distinctions between technological, organizational and human factors to be set aside in favour of the integrated design of 'orgware' (Shimada and MacDuffie 1987). This approach has been found to be highly effective in the context of integrated, IT-based systems such as CAD/CAM (Computer-Aided Design/Manufacture). Hierarchical and specialist approaches to such technologies, on the other hand, have been found seriously deficient (Winch 1983).

But whatever the functional advantages of the Japanese approach to design, the Western context of occupational specialization and interfunctional politics seems to favour more individualistic means of managing design activity. This is reflected in the adversarial implications of the term 'champion' which is applied to those individuals who co-ordinate and provide the momentum for design projects. Providing the 'champion' finds a 'sponsor' from top management, it is widely suggested that success will follow (Rothwell *et al.* 1974, Ettlie *et al.* 1984). Putting it more bluntly, 'the new idea either finds a champion or dies' (Schon 1963: 84). The implication is that successful design and innovation are largely a matter of encouraging 'champions' to emerge.

However, like many such prescriptions which relate corporate success to individual behaviour, there is a lively possibility that the correlation is tautological; the presumed cause of success is actually a consequence of it. First, many of the studies reporting innovation success actually rely on individuals reporting themselves as champions (Howell and Higgins 1990). Second, the emergence of champions, and even more of sponsors, is likely to be linked to the perceived likelihood of success of an innovation. In other words, those innovations that are seen as likely to succeed will find a champion, and perceived probable failures will not. The emergence of a champion may therefore be a self-fulfilling prophecy in many cases.

What the notion of the innovation champion does do, however, is shed light on one of the central problems of design and innovation. This problem – which reflects the subjective dimensions of knowledge noted in previous chapters – is to do with the linking of knowledge to commitment. Design activity involves not only eliciting knowledge and information from different organizational groups, but also inducing those groups and individuals to be committed to that knowledge in the sense of communicating it and acting upon it.

The problem of knowledge commitments has traditionally been associated with the scientific community, and, in particular, with the impact of industrialization on collective quality control in science (Ravetz 1971); an issue recently highlighted by the 'cold fusion' controversy. The scientific community is seen to deal with the problem through the 'invisible college': a form of occupational control which links the generation of knowledge to the distribution of rewards. The scientific community itself thus acts as 'an institutionalized system which is simultaneously a communication system and a reward system and what might be called a system for "distributing property"' (Barnes and Edge 1982: 15).

In a sense, the problems that commercial organizations face in dealing with knowledge are roughly the reverse of the threat posed by industrialization to academic science. The hierarchical structures that typify many organizations are ill-equipped to deal with the need to elicit knowledge or to generate new knowledge from all those groups involved in design activity. It is a commonplace observation of UK industry, for instance, that on the shop-floor or office-floor there is a vast reservoir of knowledge about the production process. This remains largely untapped not simply because of the problems of communicating tacit knowledge and skills, but more as a result of the social and hierarchical divisions between workers and management. Recent attempts to address this problem by introducing Japanese techniques such as Quality Circles need to be compared with a managerial tendency both to appropriate knowledge for their own exclusive use and to deny the usefulness of the knowledge possessed by other groups.

One study of a UK electrical engineering factory found that while shop-floor workers 'discuss endlessly among themselves what is wrong and ... actually carry out improvements within the scope of what lies under their control' (White 1981: 67), middle management saw such activity as a threat to their positions. Thus, a stock controller in the same factory reacted scornfully to the suggestion that shortage or wastage of components could be avoided by asking the workers about their usage: 'The day I allow myself to be dictated to by a woman assembler – or a man for that matter – is the day that I give up my job' (White 1981: 70).

The emphasis on authority and control is antithetical to the generation of knowledge commitments. The latter can only emerge where groups and individuals can freely communicate, relate to and act upon the knowledge that they develop.

THE MANAGEMENT OF KNOWLEDGE-WORKERS

This has important implications for the design circuit, in terms of the way in which different forms of knowledge can be organized and controlled. Giving recognition and rewards to knowledge implies more democratic and egalitarian structures in which status and influence derive from the possession or creation of knowledge rather than hierarchical position.

Clearly, where knowledge commitments are an important element in the design circuit there is likely to be a tension between the organization of design and the mainstream organizational hierarchy. Given the costs of developing and training in-house knowledge-workers, this may create an incentive to economize on the range of knowledges (and democratic structures) used in design. In the flexible firm approach described in Chapter 2, organizations employ different forms of control, ranging from team-based structures for those groups specially designated as in-house knowledge-workers, to tight contractual controls for external sources of expertise. When we add the role played by professional groups to these options, it is clear that the management of technical workers may involve a variety of 'modes of control' (Francis and Winstanley 1988) ranging from the professional or craft mode through managerial to market-based forms. Table 7.2 outlines the various modes together with the technical and economic conditions and forms of organization structure appropriate to each.

As Table 7.2 indicates, the tension between, on the one hand, the autonomy and knowledge commitments of knowledge-workers, and, on the other, organizational or occupational forms of regulation, produces many nuances of structure and control. Moreover, the choice is not simply between in-house versus external forms of expertise. Rather, managerial, market and professional structures of control may overlap. Thus we may find organizations developing their structures *around* the demarcations and practices of a particular craft or professional group. Alternatively, it may supplement an elaborate Taylorist structure of control with appeals to the 'professionalism' of its knowledge-workers (Tierney and Wickham 1989).

The great majority of knowledge-workers are directly employed by organizations and operate within the managerial control mode. Even within that mode, however, there is a tension between the structures deployed to achieve efficiency and productivity, and the

Table 7.2 Modes of control for technical workers

Professional/craft mode
The control of technical work is determined solely or very largely by the occupation itself.

Conditions
Services of occupational group needed at infrequent intervals. Technology and knowledge required to provide the service are not specific to a particular client. The quality of the service cannot easily be judged in the short term by the client.

Structures
Professional control structures, supported by legal and/or qualification barriers.

Professional/craft 'colonies' within organizations.

Centralized, occupation-based departments or functions – e.g. central R&D labs – headed by technical specialist 'player-managers'.

Managerial control mode
Technical work is controlled by a managerial hierarchy.

Conditions
A technologically indivisible product. Provision of a service involves investments in knowledge and technology which are specific to a particular client. Need for flexible control of technical personnel.

Structures
Decentralized, division-based technical functions.

Project-based structures, elimination of functional/discipline demarcations. Multiskilled specialists.

Internal contracts, profit centres, charging of costs.

Market control mode
Control and co-ordination of work through contracts with free agent technical workers.

Conditions
The quality of the service can be readily evaluated by the client. The service does not require investments in knowledge and technology specific to a particular client.

Structures
Disintegration; e.g. 'hiving off' the technical function as a separate company or division operating in the market-place.

Use of subcontracting and consultancy.

Source: Francis and Winstanley 1988

much looser, quasi-autonomous arrangements that foster knowledge commitments and innovation.

One solution to this problem is through the adaptation of
bureaucratic structures. The latter provide sufficient consistency
and stability to be useful frameworks of control, while simultaneously permitting a level of autonomy through an 'indulgency
pattern' (Gouldner 1954) of mutually acceptable rule evasion. The
'indulgent spaces' or 'indulgent times' permitted to knowledge-
workers may encompass a wide range of exotic behaviours as long as
certain outputs are seen to be achieved.

A now famous example of such practices is provided by the
'skunkworks' operated at the 3M company in the USA. Here,
management are prepared to countenance, indeed encourage, their
technical staff in 'bootlegging' time and resources for unofficial
projects based on their own ideas. One of 3M's most successful
innovations – 'Post-It' notes – came out of precisely such an unofficial project. Similarly, in Britain, Celltech, the biotechnology
company, allow scientists to spend up to 10 per cent of their time
on their own projects (Dodgson 1990).

KNOWLEDGE-WORKERS AND MARKET CONTROLS

As economic and technological conditions shift, however, structures
of control are likely to shift with them. The 1980s, for instance, saw
the combination of the advent of IT and structural shifts in western
economies towards the 'disorganization' (Lash and Urry 1987) of
large companies into smaller, more decentralized units. Inevitably,
such change had a range of consequences for the control of
knowledge-workers.

In both the UK (Francis and Winstanley 1988) and the USA
(Rubinstein 1985), the 1980s saw a discernible shift towards the
market mode of control for technical expertise. Although bureaucratic controls permitted occupational and discipline-based specialization to flourish, the space they gave to the knowledge
commitments of technical experts was increasingly seen as
conflicting with organizational commitments and productivity.
Whittington's (1990) study of R&D functions, for instance,
discovered that many in-house R&D labs had come to be perceived by
other business functions as 'ivory towers' or 'masonic lodges': cosy
and incestuous institutions more interested in their own esoteric
technical issues than in the problems of their host organization.

Such groups not only represented a powerful vested interest committed to what were sometimes obsolete specialist knowledges, but also represented a significant drain on resources.

The move towards a market mode of control took a variety of structural forms. Not all of them involved completely externalizing the sourcing of technical expertise, though many technical workers did make the transition from employee to subcontractor or consultant (Whittington 1991). But something of the same effect was often achieved by structural changes in the forms of authority, resources and feedback applied to the technical function. By linking these factors to customer relationships through decentralization (Whalley 1986), the creation of profit centres, and quasi-contractual relationships and competition with external suppliers (Francis and Winstanley 1988), organizations sought to induce a greater sensitivity to customers in the market-place.

This form of control is not without its costs, however. Increasing 'exposure' to the market means more than a tightening up of work practices; it also induces a shift in the skills and knowledge commitments of expert groups. On the credit side this may lead to increased knowledge of market requirements and greater skills in customer relations. But in the debit column, it has to be noted that the immediate pressures of the market and profit accountability may prove to be a poor mistress for long-term, basic research of the sort carried out by corporate research and development groups.

Although the professional controls of a scientific or technical community may have led to in-breeding and introspection, they did provide incentives for a commitment to fundamental forms of knowledge. In contrast, the market commitments elicited by a market mode of control emphasize short-term and applied projects. One research manager puts it like this: 'The bottom-line is a strong influence upon people. One is always worried that decentralization will cut people's time horizons.... Which guy focussed on next December is going to do long-term research – or even medium-term development' (Whittington 1990: 200).

IT AND THE MANAGEMENT OF KNOWLEDGE-WORKERS

While R&D workers have been increasingly subject to the logics of external market forces, computing workers have inevitably been affected by the emerging possibilities created by IT. Their experience has been instructive, for in many respects the application of IT greatly increases the transparency of the design circuit. In reducing bureaucratic distortions of communication, IT may not succeed in reducing the whole production process to neutral flows of inform-ation, but it does highlight the role played by different knowledges within that process.

This increased transparency of production provides an oppor-tunity to either shorten or extend the design circuit. On the one hand, knowledge and skills may be translated into software form and built into centralized systems. The most telling example of this approach is when an industrial robot is actually programmed by the worker that it is replacing. The worker takes the robot arm through the production task step by step, and, having recorded the various movements and functions, the robot is able to go on repeating them *ad infinitum*. The current benchmark for such uses of IT is provided by expert systems, and these are discussed in more detail below.

An alternative route, however, involves exploiting the adaptability of IT to allow the local knowledges of a wider range of user groups to shape the design and use of distributed systems. A range of tools – notably 4GLs (Fourth Generation Languages) and protoyping – has been developed to allow greater 'user involvement' in the design of IT systems.

EXPERT SYSTEMS

The development of expert systems has tended to highlight the problems rather than the benefits of translating complex forms of human knowledge into computer software. Expectations have greatly outstripped applications. Where expert systems have been successful, it has generally been for bounded problems which are capable of quantification. Moreover, far from simply eliminating experts, the expert system generally needs to be able to draw on the relatively codified knowledge of the human expert. Indeed, there is often a need for the user to be an expert. This is because the rules on which expert systems are based cannot contain the rules for when and how they should be applied (Collins 1986, Fincham 1988).

The most successful systems – notably the XCON system outlined in the case example below – have exploited IT's strengths in information processing to deal with complex, quantitative systems. In contrast, the more problematic applications have been those involving attempts to codify more interpretive and tacit forms of knowledge, with a large knowledge gap between the expert 'donor' and the end-user. Of course, it is the latter applications that are more interesting to researchers in the field. Woolgar notes that a common saying among researchers is: 'If it's useful, it isn't AI' (Woolgar 1985: 565).

Case example: The XCON expert system

XCON (an eXpert system for CONfiguration) is one of the best examples of the simple but successful class of expert systems. The system was developed by DEC to address a specific problem in the production process for their large computer systems. The company's traditional strategy of providing their customers with an almost unlimited range of technical options was creating an enormous burden of variety in production. With fifty types of central process, and 4,000 different technical options – permutations of disk drives and components, placement of components, and so on – no one in DEC could even begin to estimate the number of possible configurations from which a customer might select. Confusion and mistakes abounded, and order processing times were running at around four to six months.

The conventional solution to this problem involved some extension to the production process: at the front-end of the process, the employment of large numbers of technical editors to check customer orders before they were inputted to production; at the order completion end, FAT (Finish, Assembly and Test) plant to actually assemble and test the various configurations before they were delivered to customers. Solutions, in other words, that reflected the standard bureaucratic response to the knowledge problems created by complexity – organizational elaboration and specialization, grafting on new specialist departments to the existing division of labour. The development of XCON made such solutions obsolete, however, by offering a radically new solution to the knowledge problem. The ability to collate information on an enormous number of technical permutations meant that the control of customer orders could be more effectively done from the centre. The technical editing system, which had relied on the decentralized decision-making of experts and apprentices, was abolished. It was replaced by the central XCON system supported by human operators. A staff of forty programmers was needed to maintain the XCON knowledge-base as the changing nature of product knowledge meant that over one-half of the XCON knowledge-base had to be rewritten annually.

Dependency on the remaining technical editors was greatly reduced, however. The size of the expert editor group no longer constituted a limiting factor on the company's capacity to process configuration orders. The technical editor's job lost status and became more clerical. As one editor explained: 'It was more fun before XCON, when you had to figure out each system. You got to keep in touch with many parts of the company – engineering, sales and marketing – to know what was happening. We still do that now, but not so much' (Sviokla 1990: 137).

The development of XCON had significant strategic benefits. In particular, it allowed DEC to move towards a 'dock merge' strategy, whereby all the necessary components of a large computer system could be delivered to a customer site without previous assembly at a FAT to ensure compatibility. The cost saving was estimated at over $20 million (Leonard-Barton 1990).

USER INVOLVEMENT

Just as technological developments alone have been unable to resolve the knowledge problems associated with expert systems, the advanced computing tools of Fourth Generation Languages and prototyping have often had disappointing results in deepening the involvement of users in systems design (see Chapter 4). Setting aside the problems of organizational rigidities, such disappointments seem to highlight the problems of developing knowledge commitments within design activity.

Whatever the functional advantages of eliciting local knowledge of the use-process, users are unlikely to demonstrate the same interest in either communicating such knowledge or being committed to the design and use of IT systems based on it. This is not simply because users are reluctant to give up knowledge to another group, or because hierarchical structures fail to reward such knowledge contributions. Rather, it seems that the generation of knowledge commitments to a particular design project depends upon the users feeling that they are in control, that they have 'ownership' of the fruits of their endeavours.

According to Earl, the corollary of this point is that where user involvement is important to success, it is best achieved by transferring control from the IT function to the user department. He claims that: 'Early experimentation and R&D with new technologies seems to be more successful if it is done within appropriate user/application areas rather than under the loose wing of the IT department' (Earl 1989: 142).

The logical extension of this position has been embraced by Citibank in the USA which have moved towards every user becoming their own systems manager by simply eliminating their systems analysis function. Their user 'DIY' (do-it-yourself) approach relies on the commitment of users to becoming their own systems analysts – specifying their own requirements and managing their own projects through to completion. Back-up is provided by contract programmers (Willcocks and Mason 1987).

But Citibank have also recognized that eliminating the occupational dimension from technological design has general implications for the distribution and generation of IT knowledge. Their DIY policy is underpinned by careful selection policy and hundreds of training courses on 'Managing Technology'.

The success of such a radical redistribution of IT knowledge towards organization-specific needs remains uncertain. However, the possibility of success is usefully illuminated by comparison with some of the other proposed solutions to the need for managers to possess both generalist and IT skills. A recent report – outlined below – claims that the development of such 'hybrid managers' is becoming a critical element in competitive performance.

Hybrid managers: British Computer Society report

The British Computer Society report (1990) claims that hybrid managers – that is, 'managers possessing business understanding, technical competence and organizational knowledge and skills' (p. 3) – are needed because:

1   Information management is becoming the major value added of all professionals.
2   The compression of corporate hierarchies and elimination of middle management with the aid of information systems.
3   The globalization of production and markets requiring the development of sophisticated information systems; 'the Corporation that never sleeps is with us today – using every single

minute of the 24-hour day to achieve its global ambitions' (p. 8).

4 'Open Systems promise to reduce the cost of computing hardware and software so dramatically that IT budgets can buy any amount of product' (p. 8). The speed of conception and implementation for business applications will become the limiting factor.

The report estimates that 30 per cent of all managers will need to be hybrids by the year 2000.

Some large UK companies report significant success in developing hybrid managers; Esso UK claim that by involving hybrid managers in all IT projects, 90 per cent are now delivered on time and within budget compared to 60 per cent several years ago (*Management Today*, March 1990).

And yet, it seems debatable that managers can somehow be assembled – in Frankenstein mode – from the bits and pieces of knowledge and skill provided by job rotation and training courses. As our earlier discussion has indicated, useful knowledge involves not only the acquisition of functionally desirable forms of experience and information, but also a context in which knowledge commitments are evaluated, rewarded and nourished. In the case of specialist IT workers, this kind of context is provided by occupational specialization and a degree of interorganizational mobility. The latter permits both the acquisition and the validation of a wide range of technological expertise (Earl and Skryme 1990). Unless the organizational hierarchy is able to replicate such a framework internally there is a possibility that – as often happens currently – managers will become narrowly oriented to those forms of knowledge that garner rewards.

The US experience with CIOs (Chief Information Officers) illustrates many of the problems. This was a role that many corporations developed as a response to the pressures towards 'hybridity' in the management of IT. One recent survey, however, suggests that 'CIOs may be losing touch with their technological roots as well as with the business side' (Datamation, 15 August 1990: 31). Worse, the Datamation survey indicates that the job-hopping encouraged by labour-market pressures – an average 25 per cent increase in salary with each job move – is sustaining a professional rather than an organizational orientation to the job.

SUMMARY

The development of hybrid managers or CIOs indicates the extent to which the formation of knowledge is dependent upon the operation of broader social and economic institutions. One example of such dependency is the link between an organization's ability to pursue human resource strategies based on employee training and development and the operation of the broader labour-market. It can be argued, for instance (Streeck 1989), that unregulated labour-markets encouraging labour mobility will tend to undermine such strategies. Employers may be unwilling to invest in the skills and knowledge of employees who, as they become more skilled, are more likely to be poached by competitors. Conversely, the existence of an unregulated, external labour-market is likely to reinforce the tendency towards the 'buying-in' of specific skills as and when needed.

Such factors have implications for an organization's ability to reshape its design circuit to exploit the opportunities created by technological and economic developments. At the very least, given the long-term aspects of learning and knowledge acquisition, they highlight the significance attached to current product-market and human resource strategies. The latter help to determine both the future shape of the design circuit and the knowledges that it is likely to contain.

Chapter 8

Technology, management and organization
Some conclusions

INTRODUCTION

We have already seen how the interaction between technology and organization is influenced by the 'design circuit' of the distribution of knowledges between different groups. Similarly, in earlier chapters we described the role of such groups in sustaining the circuits of power and meaning that shaped the technology process. In particular, we noted the persistence of long-established models such as Taylorism and Fordism which both reflect and, at an ideological level, continue to shape the interests and actions of management and the expert groups within it. However, in discussing moves towards Flexibility, we also noted the possible erosion of the political and technological underpinnings for such models, and for the distribution of knowledge that they perpetuate. The problems of strategic control posed by changes in product-markets and technology are seen as opening up the possibility of proactive management action to reshape the design and application of technology.

In this final chapter, therefore, we will focus specifically on the extent to which management are able to direct the development of technology and organization in a purposeful and strategic manner. On the one hand, and particularly since the advent of IT, there are prima-facie arguments (e.g. Frohman 1985) that competitive forces are making a strategic approach to technology an imperative feature of organizational performance. On the other is the view, which has surfaced a number of times in previous chapters, that technology and organization are closely intertwined and their unfolding uncertain and contradictory, and that consequently hierarchical control alone does not equate with the ability to shape the technology process.

Although the arguments about the strategic control of technology are as normative as they are analytic, reviewing their feasibility in different contexts can tell us a great deal about the relationship between management and technology in general. Much has been made, for instance, of the market and competitive advantages of using technology strategically. Technology, it is said, can help to 'lock in' customers and raise barriers to entry to rivals (Porter and Millar 1985). Certainly, there are important incentives for organizations to shape technological design proactively, and there are some notable examples of successful outcomes being achieved. Green (1990), for instance, in describing the case of Abbott Pharmaceuticals outlined below, highlights the extent to which companies can actively and lucratively shape the competitive rules of the game and create a 'market space' for themselves through technology.

Influencing the rules of the game: Abbott Pharmaceuticals

The example of the Abbott pharmaceuticals corporation in the USA and their development of the 'immunoassay' market shows how one company can create and dominate a 'market space'. Immunoassays involve the use of antibodies to detect the presence of bacteria and viruses in blood or urine. Up to the 1970s this technology had been largely developed and controlled by hospital laboratories, and although there had been a limited market for the necessary instrumentation, most of the technology was incorporated in the skills and knowledge of hospital lab technicians.

In this context, Abbott sought to open up a market space for their new immunoassay product through a clever marketing strategy. Abbott offered hospitals a complete immunoassay system, i.e. one that incorporated both the analysing instrumentation and the packaged chemicals required for the tests. In addition, they offered a 24-hour service guarantee. Finally, and most importantly, 'by selling analysers exclusively made for itself, Abbott locked buyers into the purchase of Abbott reagents and service contracts' (Green 1990: 12).

Once dependent on the Abbott product for their immunoassay technology, hospital labs ceded control of developments in this area and became dependent on Abbott's R&D for future advances in immunology. The rules of the game were transformed and a new market created as all Abbott's major competitors were forced to follow their strategy of offering complete machine-reagent systems.

But the ability to achieve such market success is not simply a matter of rationally working through the strategic possibilities of a particular technology. It also depends upon the organization's ability to manipulate the invisible web of interests and perceptions that ties it to suppliers, rivals and customers.

TECHNICAL STANDARDS

A useful example of the way in which technology itself affects this web of interdependency is provided by the issue of 'technical standards', which are a key competitive area in many technology markets. Standards are an important and necessary feature of markets where questions of technical compatibility (between pieces of hardware or between hardware and software) are significant. If a dominant standard fails to emerge in such a market, there are significant disincentives to purchasing a particular product for fear of obsolescence or incompatibility. As a result, the market itself may simply never take off.

A classic example of the importance of standards is provided by the VCR (video cassette recorder). Phillips in Europe originated the innovation but it was Japanese firms who established the VHS standard and who dominated the market thereafter. A decade or so later, Phillips showed that they had learned this lesson well by collaborating with Sony of Japan on the development of the CD (compact disc) player. The collaboration ensured that the technical standards of the new product would be at least partially defined by Phillips, allowing them to establish a powerful position in this new market.

The Phillips example demonstrates the extent to which market success in technology industries has come to depend upon interfirm collaboration more than on 'dog eat dog' competitive tactics. Indeed, sometimes market success can only be achieved by giving it away. That is, in order to establish the dominance of its own standard, a firm must open it up to competing firms to widen the market-base of acceptability, and discourage the promotion of alternative standards. IBM, for example, were able to establish the dominance of their PC (Personal Computer) MS-DOS standard in personal computing by encouraging large numbers of software houses to write software for it. At the same time, however, the sales performance of IBM's own machines was undermined by the influx of low-cost PC clones into the open market that the new standard created.

THE DIFFUSION OF TECHNOLOGY

The issue of technical standards illustrates the importance of the distribution of technological knowledge between manufacturers, and the advantages to be gained from the collaborative sharing of that knowledge in terms of creating new markets. Equally important, however, as the Abbott case demonstrates, are the pre-existing technological knowledges and understandings of consumers themselves. The interaction between such pre-existing knowledge and the knowledges associated with a new technology product will help to determine the latter's acceptability and diffusion. Neglecting the importance of such factors, however, can lead to the kind of abortive developments that are typified by the example of screen-based home banking in the USA.

In the late 1970s, one consulting firm predicted that screen-based home banking would be accepted by 45 per cent of all US households by 1985. Following the trend of expectations, huge sums of money – something of the order of $600 million – were invested in this technology by US financial institutions. Yet by the late 1980s only 100,000 people out of a US population of around 200 million had actually adopted home banking technology.

The debacle of home banking in the USA provides a classic example of the constraints on the diffusion of both technological knowledge and artefacts. Rogers's (1962) analysis of diffusion emphasizes the importance of individual perceptions and understanding of a new technology in shaping its acceptance. When applied to the home banking example it usefully demonstrates some of the reasons for its failure in the USA (see Table 8.1).

The problem of constraints on diffusion directs attention to the issue of consumer education and attitudes in relation to new technology. But such external conditions can also be seen as mirroring the internal problems of knowledge and expertise within the organization. The openness of organizations to new technology is similarly conditioned by the distribution of pre-existing knowledges and attitudes. Thus, the forging of connections between technology and strategy is as much to do with the internal organization of management as it is with external strategizing of the market. Of particular importance, as we noted earlier of the design circuit, is the extent to which technological knowledge tends to be compartmentalized into managerial and technical specialisms.

Table 8.1 Constraints on the diffusion of home banking in the USA and UK

Feature	Constraint
Relative advantage (the degree to which an innovation is superior to the ideas it supersedes)	Initial start-up cost. $900 in USA and over £100 for Prestel in the UK
Compatibility (the degree to which an innovation is consistent with existing values and past experience)	Lack of a dominant technical standard and fears of obsolescence
Complexity (the degree to which an innovation is difficult to understand or use)	Technophobia, perceived complexity
Divisibility (the degree to which an innovation may be tried on a limited basis)	Limited divisibility given infrastructural connections
Communicability (the degree to which the results of an innovation may be communicated to others)	Complex decision-making process Concern over security and ease of use

Source: Dover 1988

TECHNOLOGICAL AND MANAGERIAL KNOWLEDGE

The strategic implications of the generally uneven distribution of technological and managerial knowledge are most evident at board-room level. This is not just a question of general management education and training, but of the absence or presence of specific forms of knowledge at senior levels. The relative paucity of engineering knowledge at the highest level of UK companies, for instance, can be contrasted with the superabundance of finance and accounting expertise. This in itself may have implications for the kind of strategies and structures that UK organizations evolve. Thus, even at an engineering company like GEC, financial controls predominate. The *Financial Times* (2 October 1987) reports the response of one senior executive at GEC, when asked how many years a manager could fail to meet targets before being sacked: 'How many years? You mean how many months. He might last for six months or he might not.'

Similarly, in UK industry generally, the strategic use of IT is clearly constrained by the general lack of IT knowledge at board level (EOSYS 1986). Such factors draw attention to the role played by the division of labour, both vertically and horizontally, within management, in determining the kinds of knowledge that influence strategy and decision-making. This suggests that changes in both structure and decision-making criteria (McFarlan 1984) may be preconditions to the formation of a usable strategic knowledge of technology.

However, in saying this a crucial caveat must be entered. Given that existing structures and criteria are, as we have noted, reflections of deeply embedded and implicit knowledge, and are therefore quite as 'rational' as – if not more than – any putative replacement, it seems unlikely that any such change will occur through revelation alone. A much more likely scenario, in fact, is one in which the development of new ways of thinking about technology emerges slowly and painfully out of a process of organizational learning. One example of such a learning process is provided by the case example of the Bank of Scotland below.

Case example: Bank of Scotland

By the mid-1980s, the Bank of Scotland had established themselves as a highly innovative and strategically minded user of IT. This was reflected in a range of innovations which included a new Branch Information Network and a screen-based 'Home and Office Banking System'.

This successful and energetic exploitation of IT had its roots, however, in a learning process which had been stimulated by failure. In the late 1970s, the Bank had decided to stand back from adopting ATMs (Automated Teller Machines) on the grounds that they were unable to justify themselves in terms of the financial benefits of automating branch clerical labour. However, their major rival, the Royal Bank of Scotland, had pressed ahead and had obtained significant marketing advantages from developing an extensive network of ATMs.

In the aftermath of this competitive setback, the Bank reviewed the decision-making processes and structures that had informed the original decision on ATMs. An Automation Planning Department was established to provide a strategic overview of IT, in order to complement the predominantly specialist views emanating from the existing Computer Services and Management Services divisions. Later, the head of this department took over as General Manager of a reconstituted Management Services Division – incorporating Management and Computer Services – and from this base helped to spearhead many of the 1980s' innovations.

Source: Scarbrough and Lannon (1988)

COMPETENCES AND COMPETENCE TRAPS

The importance of organizational and technological learning to the formation of strategy is increasingly being recognized (Whipp and Clark 1986). One consequence is that the equation of strategy with rational manoeuvre and adaptation is being challenged by a notion of strategy as an 'architecture that guides competence building' (Pralahad and Hamel 1990: 91). Thus, attention is drawn to the way in which product-market success increasingly depends upon the focused acquisition of technological knowledge through interfirm collaboration. The building of competences rather than the financial management of portfolios is seen as the critical strategic task. Much of the thinking behind this view is based upon the success of Japanese organizations in developing interrelated technological competences. The NEC corporation, for instance, is said to have entered into over 100 strategic alliances between 1980 and 1987 'aimed at building competences rapidly and at low cost' (Pralahad and Hamel 1990: 80).

However, while this normative view rightly stresses the strategic implications of technological learning and competence, it is important to acknowledge the 'downside' of these phenomena. As we noted earlier in discussing the design circuit, the learning process generates embedded and implicit forms of knowledge which empower but also constrain strategy and management. At one level, this may mean that the organizing knowledge centred on existing technologies delimits the use-process for new technologies. The use of new IT systems in the 1980s, for instance, was often dominated by this embedded organizing knowledge. A typical example is provided by a recent study of office automation (Webster 1990) which describes the influence of this organizational *status quo* on the use of IT. On a positive note, one consequence of the pre-existing organizing knowledge was that the 'de-skilling' potential of the new systems was not explored. By and large, personal secretaries retained their diversity of tasks and relative autonomy. But by the same token, the work of the women in the typing pool remained as routine and impersonal under the new technology as it had under the old.

Similarly, in the manufacturing context, McLoughlin (1990) notes the tendency for the adoption and use of CAD (Computer-Aided Design) technology to be simply 'grafted' onto the existing organization structure as a specialist function. And at shop-floor level, Cummings and Blumberg (1987) claim that the 'ineffective' use of advanced manufacturing technology is often linked to a 'failure' to redesign the work process and context when new technological systems are implemented. The pre-existing task structures and functions are seen as preventing the emergence of more appropriate organizational forms such as self-regulating groups.

The durability of the organizing dimension to technological change – which perhaps makes talk of 'failure' and 'ineffectiveness' slightly spurious – is indicated by the parallels such examples reveal between the adoption of IT in the 1980s and electric power in the 1880s, as outlined below.

Case example: Electric power in the late nineteenth century

The key inventions for electric power were made in the 1860s–1880s. The 1880s and 1890s saw the establishment of generating and transmission systems. In retrospect, the unit drive of electric motors can be seen to have possessed a number of advantages, notably: greater flexibility in factory layout because machines no longer had to be placed in line with shafts; big capital savings in floor space; and increases in the flexibility and adaptability of the production system through the use of portable power tools. However, for many years these potential gains were not achieved because the new electric motors were installed in a factory layout still shaped by the elaborate shaft and belt transmission systems needed for steam engines. Given the accumulation of detailed knowledge of shaft and belt transmission, many companies found it easier to retain existing factory layouts and simply replace the steam engine with the electric motor.

Source: Devine (1983)

But if embedded organizing and technological knowledge is an influence on the use of new machines at shop-floor level then it is even more important at a strategic level. March and Sproull describe the kind of 'competence traps' that may result: 'Increasing competence within an existing technology (or other knowledge structure such as a scientific paradigm) makes it difficult to shift to new and potentially better technologies. In effect, learning drives out the experimentation on which it depends' (March and Sproull 1990: 161).

The danger is greatest in established industries which are challenged by some radically new technology. As Foster (1986) has described in his analysis of the 'S curve' of technological development (see Figure 8.1), the incentive to adopt a new and competing technology in the early stages of its evolution is by no means clear-cut. The new technology will typically perform less well than the established technology at this stage. Where the new technology does perform well it may only be for limited applications or specific market segments. The transistor, for example, found early application in hearing aids and pocket radios, but not in radar systems and television (Cooper and Schendel 1982).

1   Learning curves involve an investment in expertise specific to a particular technology.
2   All technologies have limitations on their performance, leading to diminishing returns.
3   Existing body of expertise will define problems of performance in terms of need to refine existing technology.

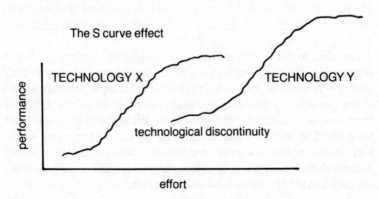

Figure 8.1 The 'S curve'
Source: Foster 1986

And even when the economic rationale for making the switch becomes more pressing, it may be compromised by the hold that established knowledge commitments can exert on perceptions and decisions. Rather than switching to the new technology, many firms believe or 'know' that they can continue to outperform it by squeezing out higher performance from their existing technology. This leads to what can be termed the 'sailing ship effect' which is described below.

The 'sailing ship effect'

The first phase of the contest between sail and steam in the nine-teenth century was characterized by the unreliability and poor performance of the new steam-engined vessels. Although such experimental failures were in some senses inevitable, the sailing-ship designers and builders of the day were quite happy to take them as evidence of the inherent unreliability of steam power. Consequently, further improvements in the new steam technology only stimulated sailing-ship designers into extracting ever-higher levels of performance from sail. This involved a progressive increase in the numbers of masts and the area of sail carried by each ship. In the process, however, sailing-ship designers had to sacrifice manoeuvrability and stability. They finally overreached themselves in the design of what was to be the last 'state of the art' cargo sailing ship. The seven-masted *Thomas Lawson* capsized at anchor off the Scilly Isles in 1907.

Source: Foster (1986)

A fate akin to that of the sailing-ship companies continues to befall organizations whose technological competences dominate their strategy. Braun and Macdonald (1978), for example, see this as one of the reasons for the relative competitive failure of the UK and European semiconductor industry in the 1960s. Unlike its American rival, the UK industry was dominated by large companies which had their origins in valve technology. An industrial scientist described the frustrations of attempting to introduce new techno-logical knowledge into such a company thus:

The production division will carry on doing what they have done and do it a little better because they have large technological resources and if you come up with a new invention they will be against it. These fellows will say, 'Oh well, we can probably do it by tweeking our existing processes.' And they do until they make a better transistor than you can make with a revolutionary idea, because in the beginning you aren't high up on the learning curve. So you don't win and it takes a long time, until everybody is really screaming, that the new will take over.

(Quoted in Braun and Macdonald 1978: 154)

When companies do succeed in switching from one major technology to another, they often do so belatedly and at a high cost. In May 1971, NCR (National Cash Registers) had to announce that $140 million worth of newly designed cash registers were impossible to sell and would be written off. The machines could not compete with the new, cheaper, easier to use electronic machines. In the following months, the company succeeded in transforming itself into an electronics machine company, but at a price: many thousands of workers were laid off and the chief executive was fired.

BEYOND STRATEGY?

The speed of product-market change and the explosion of technological possibilities can all be seen as legitimating or stimulating the formation of a strategic management of technology and organization. However, a number of the factors described above suggest that the notion of strategy may be an ineffective way of knowing about and acting on the technology and organizing processes. This is not only because the political and cultural aspects of organizations blunt the implementation of strategy, or encourage the kind of *ad hoc* or piecemeal uses of technology that we noted earlier. Rather, it is to do with, for instance, the ability of strategists to understand, and capture, the flows of knowledge out of which new technologies emerge. The attempt to organize networks of information, knowledge and technology around a strategic focus must at best inhibit the range of technologies in which the organization can be engaged. At a high technology company like STC, for instance, the exercise of 'strategic planning' means that the corporate centre must be knowledgeable about each of their – limited – range of businesses. As an executive at STC put it: 'For our approach to succeed, the centre must understand the businesses well to be able to challenge the management of companies on the quality of their proposals' (Goold and Campbell 1987: 44).

At worst, however, such an approach may encourage the kind of enmeshing of technological knowledge with organizational hierarchy that leads to the 'competence traps' noted above. Equally importantly, the attempt to construct a 'focal organization' around the strategy-environment paradigm may inhibit the development of the transitory interorganizational relationships – the 'alliances' developed by NEC, for instance – that are better able to mobilize technological knowledge and resources. In effect, where 'strategic sovereignty' means the single-minded pursuit of organizational interests against all comers, it is likely to squeeze out the possibilities for local, extra-organizational relationships and flows of knowledge.

The consequences of such an approach are illustrated by the convention of the 'multisourcing' of components which was a typical feature of Western, especially US, car producers up to the mid-1980s. The aim of 'multisourcing' was to play off one component supplier against another, and thereby, in theory, achieve the lowest costs and highest quality. However, the arm's-length relationship that this strategy created inevitably constricted the exchange of knowledge and information between car firm and components supplier. As a result, technological development was retarded as innovations were not readily transmitted between one supplier and another, or even between producer and supplier, lest an 'uncompetitive' advantage resulted (Womack *et al.* 1990).

The Japanese, of course, seem to organize such relationships more effectively. But, in the UK too, the case example of the garment retailers Marks and Spencer provides an instance of the management of interorganizational relationships for mutual advantage.

Case example: Marks and Spencer

Marks and Spencer (M & S) have been described as a 'manufacturer without factories'. The M & S relationship with their suppliers is one of detailed, long-term involvement. Of 800 suppliers, 150 have manufactured goods for M & S for more than 25 years.

Within such a relationship, M & S have developed a 'technological approach to retailing'. That is to say, they were able to stimulate and direct technical design and innovation through supplier relationships, rather than through vertical integration into manufacturing themselves. In terms of product innovation, for instance, M & S set up their own laboratories to develop new fabrics and dyes, but then passed on the knowledge gained to a manufacturer for production purposes. M & S were also prepared to underwrite the cost of their suppliers' own research and product development which was often beyond the means of the individual manufacturer.

Similarly, process innovation and design could be influenced by clearly defining standards of quality, and thereby effectively setting technical standards, and encouraging best practice techniques.

Marks and Spencer have been able to develop this extensive, but highly effective, circuit of technological design through a long-term relationship with suppliers. Supporting that relationship, and giving Marks and Spencer a dominant position within it, has been the company's own market success as a retailer. This has given it the ability to offer large, long-term orders to suppliers.

Source: Braham (1985)

In the context of technological development, the trends towards the 'disorganization' (Lash and Urry 1987) of industry, which were noted in Chapter 1, highlight the increasing importance of the kind of interorganizational relationships revealed by the Marks and Spencer case. In the terms used in previous chapters, such arrangements allow the elaboration of a 'design circuit' which transcends structural boundaries.

Moreover, when the widespread application of IT is added to the importance of such interorganizational links, it suggests a different view of organizational forms in general. Rather than viewing organization structures as the product of overarching strategies, such forms can be understood, and perhaps increasingly implemented, simply as 'modes of organizing'. In this guise, structures reflect, and not inhibit, the underlying flows of knowledge and information involved in technological design and innovation.

Child (1987a) outlines some of the major organizing modes – described in the example overleaf – based on clusters of relationships between knowledge, information and control. Such modes emerge first out of the extent to which knowledge has been codified (Boisot 1986). Broadly, the more intensive and unstructured the knowledge involved in a design activity, the greater the need to control it through a hierarchical organization. Conversely, where information is highly codified, the design circuit can be constructed through more market-based arrangements.

Second, they depend on the diffusion of information: the extent to which information about the design activity is widely dispersed and communicated. The more widely diffused such information is, the more possible it is to control the design activity through market-based structures.

Modes of organizing

Integrated hierarchical mode

This 'mode of organizing' is most appropriate where design activity involves high levels of co-ordination and transfer of detailed and tacit knowledge. Classic examples include the small-batch, high-variety engineering shop.

Semi-hierarchical mode: multi-divisional/holding company

Where the degree of overlap between markets and technologies is low, design activity can be decentralized. Control of design activity can be done through a limited range of highly codified forms of knowledge and information, typically financial controls.

Holding companies exploit this clear-cut division of knowledge by putting their subsidiaries on an arm's-length, quasi-market basis.

Co-contracting mode: joint ventures

Where some elements of design benefit from the pooling of knowledge and R&D investments, while others can be carried out on a modular basis.

An example would be the Airbus Industrie consortium, the makers of the European airbus, which by modularizing airplane design (engines to one country, wings to another) have been able to achieve a high level of decentralization – sufficient at least to meet the needs of national prestige – while providing the necessary degree of co-ordination and control for the overall design effort.

Co-ordinated contracting mode: contractor and subcontractors

Again the benefits of this arrangement – at least when carried out on a long-term basis – are its ability to offer some degree of hierarchical control and knowledge transfer, while retaining the control advantages of a market relationship with the subcontractors. This mode is a characteristic feature of Japanese industry, where production techniques such as Just-In-Time depend on close and long-term relationships between component suppliers and auto manufacturers.

Adapted from Child (1987a)

SUMMARY AND CONCLUSIONS

This chapter has served to consolidate our general theme that the relationship between technology and organization is neither one of 'impacts' nor of 'choice' *per se*. Rather, technology and organization are closely intertwined through flows of knowledge and ideas which transcend the individual organization but which find expression in, and are reinforced by, political interests and agendas at the organization level.

Even at a strategic level, the flow and distribution of technological knowledge within management, and between the organization, customers and suppliers, seems to shape, not merely reflect, the implementation of corporate strategy. We have noted the extent to which strategies may be so embedded in the knowledges of an existing 'design circuit' that technological competences become *de facto* strategic commitments. We have also noted the problems and contradictions in attempts to capture technology within a strategic framework: specifically, the importance of interorganizational networks and the detrimental effect of attempting to centralize and control all such relationships through a strategic focus. This seems to reinforce our general view that technology and organization are not ontologically separate categories, but rather mirror reflections of a mutual interchange of knowledge, meanings and political interests. The sense of management, or even organizations, being able to stand outside or make abstract 'choices' about the technology process, neglects the extent to which managers' or anyone else's ability to choose is a function of the knowledge and meanings that they mobilize within that process.

Thus, if, in time-honoured fashion, we conclude by looking ahead to the future development of the technology–organization relationship, we see this nexus being defined not so much by technological hardware as by the underlying meanings and knowledge that different groups pursue within the technology process. The further application of IT systems, for instance, may continue to be shaped by Fordist notions of control and concentrated expertise. Conversely, the emerging possibilities of IT may indeed change the way in which organizations operate. Not, however, in the conventional sense of encouraging centralization or decentralization; as information can flow both ways, IT could equally be used for either purpose. Rather, to return to the theme with which we introduced this book, over time the diffusion of IT as hardware and expertise may subtly change management's way of 'knowing' about the organization. It may shift the basis of management from paper-bound bureaucracy to the reading of an 'electronic text' (Zuboff 1988). As a result, rigidities and boundaries may be highlighted, and new 'modes of organizing' revealed. Thus, knowledge – that is, new ways of seeing the world – reveals itself as still the most important tool for acting on it.

References

Abernathy, W.J. (1978) *The Productivity Dilemma: Roadblock to Innovation in the Automobile Industry*, Baltimore: Johns Hopkins University Press.

Ackoff, R.L. (1983) 'Beyond prediction and control', *Journal of Management Studies* 20: 59–69.

Albury, D. and Schwartz, J. (1982) *Partial Progress: The Politics of Science and Technology*, London: Pluto Press.

Armstrong, P. (1985) 'Changing management control strategies: the role of competition between accountancy and other organisational professions', *Accounting, Organizations and Society* 10: 129–48.

——— (1988) 'Labour and monopoly capital', in R. Hyman and W. Streeck (eds) *New Technology and Industrial Relations*, Oxford: Basil Blackwell.

Atkinson, J. (1984) 'Manpower strategies for flexible organisations', *Personnel Management*, August, 28–31.

Attewell, P. (1987) 'The deskilling controversy', *Work and Occupations* 14: 323–46.

Babbage, C. (1832) *The Economy of Machines and Manufactures*, New York: Kelley.

Bachrach, P. and Baratz, M.S. (1962) 'Two faces of power', *American Political Science Review* 30: 224–41.

Baldamus, W. (1961) *Efficiency and Effort*, London: Tavistock.

Barley, S.R. (1990) 'The alignment of technology and structure through roles and networks', *Administrative Science Quarterly* 35: 61–103.

Barnes, B. and Edge, D. (1982) *Science in Context: Readings in the Sociology of Science*, Milton Keynes: Open University Press.

Batstone, E., Gourlay, S., Levie, H. and Moore, R. (1987) *New Technology and the Process of Labour Regulation*, Oxford: Clarendon Press.

Beirne, M. and Ramsay, H. (1988) 'Computer redesign and labour process theory: towards a critical appraisal', in D. Knights and H. Willmott (eds) *New Technology and the Labour Process*, London: Macmillan.

Berniker, E. (1987) 'Understanding technical systems', presented paper, Symposium on Management Training Programs: Implications of New Technologies, Geneva, Switzerland.

Bijker, W.E., Hughes, T. and Pinch, T.J. (eds) (1987) *The Social Construction of Technological Systems*, Cambridge, Mass. and London: MIT Press.

Blauner, R. (1964) *Alienation and Freedom: The Factory Worker and His Industry*, Chicago: University of Chicago Press.

Bloomfield, B.P. (1989) 'On speaking about computing', *Sociology* 23: 409–26.

Boguslaw, R. (1965) *The New Utopians: A Study of Systems Design and Social Change*, New York: Prentice-Hall.

Boisot, M.H. (1986) 'Markets and hierarchies in a cultural perspective', *Organization Studies* 7: 135–58.

Braham, P. (1985) 'Marks and Spencer: a technological approach to retailing', in E. Rhodes and D. Wield (eds) *Implementing New Technologies: Choice, Decision and Change in Manufacturing*, Oxford: Basil Blackwell.

Braun, E. and Macdonald, S. (1978) *Revolution in Miniature: The History and Impact of Semiconductor Electronics*, Cambridge: Cambridge University Press.

Braverman, H. (1974) *Labor and Monopoly Capital*, New York: Monthly Review Press.

British Computer Society (1990) *From Potential to Reality: A Report by the British Computer Society Task Group on Hybrids*, London: British Computer Society Publications.

Bronowski, J. (1973) *The Ascent of Man*, London: BBC Publications.

Buchanan, D.A. (1983) 'Technological imperatives and strategic choice', in G. Winch (ed.) *Information Technology in Manufacturing Processes*, London: Rossendale.

—— and Boddy, D. (1983) *Organizations in the Computer Age*, Aldershot: Gower.

—— and Huczynski, A.A. (1985) *Organizational Behaviour*, London: Prentice-Hall.

Burawoy, M. (1985) *The Politics of Production*, London: Verso Press.

Burns T. and Stalker G.M. (1961) *The Management of Innovation*, London: Tavistock Publications.

Burrell, G. (1983) 'Systems thinking, systems practice: a review', *Journal of Applied Systems Analysis* 10: 105–8.

Bylinsky, G. (1986) 'A breakthrough in automating the assembly-line', *Fortune*, 26 May.

Checkland, P.B. (1981) *Systems Thinking, Systems Practice*, Chichester: Wiley.

Child, J. (1972) 'Organisation structure, environment and performance: the role of strategic choice', *Sociology* 6: 1–22.

—— (1985) 'Managerial strategies, new technology and the labour process', in D. Knights, H. Willmott and D. Collinson (eds) *Job Redesign: Critical Perspectives on the Labour Process*, Aldershot: Gower.

—— (1987a) 'Information technology, organization and the response to strategic challenges', *California Management Review* 30: 33–49.

—— (1987b) 'Organisational design for advanced manufacturing technology', in T.D. Wall, C.W. Clegg and N.J. Kemp (eds) *The Human Side of Advanced Manufacturing Technology*, Chichester: Wiley.

—— , Loveridge, R., Harvey, J. and Spencer, A. (1984) 'Microelectronics and the quality of employment in services', in P. Marstrand (ed.) *New Technology and the Future of Work and Skills*, London: Frances Pinter.

—— and Smith, C. (1987) 'The context and process of organizational transformation – Cadbury Limited in its sector', *Journal of Management Studies* 24: 565–93.

Clark, J., McLoughlin, I., Rose, H. and King, R. (1988) *The Process of Technological Change*, Cambridge: Cambridge University Press.

Clark, P.A. and Starkey, K. (1988) *Organization Transitions and Innovation-Design*, London: Pinter Publishers.

—— and Staunton, N. (1989) *Innovation in Technology and Organization*, London: Routledge.

Clegg, C.W. and Corbett, J.M. (1987) 'Research and development into "humanizing" advanced manufacturing technology', in T.D. Wall, C.W. Clegg and N.J. Kemp (eds) *The Human Side of Advanced Manufacturing Technology*, Chichester: Wiley.

Cockburn, C. (1983) *Brothers: Male Dominance and Technological Change*, London: Pluto Press.

—— (1985) *Machinery of Dominance: Women, Men and Technical Know-how*, London: Pluto Press.

Collard, R. and Dale, B. (1989) 'Quality circles', in K. Sisson (ed.) *Personnel Management in Britain*, Oxford: Basil Blackwell.

Collingridge, D. (1980) *The Social Control of Technology*, Milton Keynes: Open University Press.

Collins, H.M. (1986) 'Expert systems, artificial intelligence and the co-ordinates of action', *Conference on Technology and Social Change*, University of Edinburgh, June.

—— and Pinch, T.J. (1982) *Frames of Meaning: The Social Construction of Extraordinary Science*, London: Routledge & Kegan Paul.

Cook, J.D., Hepworth. S., Wall, T.D. and Warr, P. (1982) *The Experience of Work*, London: Academic Press.

Cooley, M.J.E. (1988) *Architect or Bee?*, London: Hogarth Press.

Cooper, A.C. and Schendel, D. (1982) 'Strategic responses to technological threats', in M.L. Tushman and W.L. Moore (eds) *Readings in the Management of Innovation*, London: Pitman.

Corbett, J.M. (1987) 'A psychological study of advanced manufacturing technology: the concept of coupling', *Behaviour and Information Technology* 6: 441–55.

—— (1989) 'Automate or innervate: the role of knowledge in advanced manufacturing systems', *AI and Society* 3: 198–208.

——, Rasmussen, L.B. and Rauner F. (1991) *Crossing the Border: The Social and Engineering Design of Computer Integrated Manufacturing Systems*, London: Springer Verlag.

Council for Science and Society (1981) *New Technology: Society, Employment and Skill*, London: Blackrose Press.

Cressey, P. (1985) *Consolidated Report on the Role of Parties in the Introduction of New Technology*, Dublin: European Foundation Publications.

—— (1987) 'New technology: an overview of regulation', *European Industrial Relations Review* 157: 9–16.

Cummings, T. and Blumberg, M. (1987) 'Advanced manufacturing technology and work design', in T.D. Wall, C.W. Clegg and N.J. Kemp (eds) *The Human Side of Advanced Manufacturing Technology*, Chichester: Wiley.

Daniel, W.W. (1987) *Workplace Industrial Relations and Technical Change*,

London: Frances Pinter/Policy Studies Institute.

—— and Hogarth T. (1990) 'Worker support for technical change', *New Technology, Work and Employment* 5: 85–93.

Davis, L.E., Canter, R.R. and Hoffman, J. (1955) 'Current job design criteria', *Journal of Industrial Engineering* 6: 5–11.

DeMarco, T. (1979) *Structured Systems Specification*, New York: Yourdon.

Devine, W. (1983) 'From shafts to wires: historical perspective on electrification', *Journal of Economic History* 43: 347–72.

Dickson, D. (1974) *Alternative Technology and the Politics of Technical Change*, Glasgow: Collins/Fontana.

Dickson, T., McLachlan, H.V., Prior, P. and Swales, K. (1988) 'Big Blue and the unions: IBM, individualism and trade union strategy', *Work, Employment and Society* 2 (4): 506–20.

Dodgson, M. (1990) 'Strategy and technological learning: an interdisciplinary microstudy', paper presented at the International Conference on Firm Strategy and Technological Change, Manchester.

Dover, P.A. (1988) 'The effect of technology selection on consumer adoption of in-home computerised banking', *International Journal of Bank Marketing* 2: 31–7.

Drucker, P. (1988) 'The coming of the new organization', *Harvard Business Review*, Jan.–Feb., 45–53.

Dunford, R. and McGraw, P. (1986) 'Quality circles or quality circus? Labour process theory and the question of quality circle programmes', paper presented at the Aston-UMIST Conference on the Labour Process.

Earl, M.J. (1989) *Management Strategies for Information Technology*, London: Prentice Hall.

—— and Skryme, D.J. (1990) *'Hybrid Managers': What Do We Know About Them?'*, Working Paper RDP90/6, Oxford Institute of Information Management, Oxford: Templeton College.

Edwards, P.N. (1985) 'The closed world: computers and the politics of discourse', unpublished Ph.D. dissertation, Silicon Valley Research Group, University of California, Santa Cruz.

—— (1988) 'The closed world: systems thinking and military discourse in post World War Two historical consciousness', *AI and Society* 2: 94–112.

Edwards, R. (1979) *Contested Terrain*, London: Heinemann.

Ehn, P. (1988) *Work Oriented Design of Computer Artifacts*, Stockholm: Arbetslivscentrum.

——, Kyng, M. and Sundblat, Y. (1981) 'Training, technology and product from the quality of work perspective', *UTOPIA Report Number 2*, Stockholm: Arbetslivscentrum.

Ellul, J. (1974) *The Technological Society*, New York: Vintage Books.

Emery, F. (1978) 'The assembly-line: its logic and our future', *International Studies of Management and Organization* 8: 82–100.

EOSYS (1986) *Top Executives and Information Technology: Disappointed Expectations*, Slough: EOSYS Ltd.

Ettlie, J.E., Bridges, W.P. and O'Keefe, R.D. (1984) 'Organization strategy and structural differences for radical versus incremental innovation', *Management Science* 30: 682–95.

Fallows, J. (1985) 'The American Army and the M-16 rifle', in D.

MacKenzie and J. Wacjman (eds) *The Social Shaping of Technology*, Milton Keynes: Open University Press.

Fincham, R. (1988) 'Organizational implications of expert systems', mimeo (personal communication).

Fleck, J. (1987) *Innofusion or Diffusation: The Nature of Technological Development in Robotics*, University of Edinburgh, Dept. of Business Studies, Working Paper Series 87/9.

Forester, T. (ed.) (1980) *The Microelectronics Revolution*, Oxford: Basil Blackwell.

Foster, R.N. (1986) *Innovation: The Attacker's Advantage*, London: Macmillan.

Foucault, M. (1977) *Discipline and Punish: The Birth of the Prison*, London: Allen Lane.

Fox, A. (1974) *Beyond Contract: Work, Power and Trust Relations*, London: Faber & Faber.

——— (1980) 'The meaning of work', in G. Esland and G. Salaman (eds) *The Politics of Work and Occupations*, Milton Keynes: Open University Press.

Francis, A. and Winstanley, D. (1988) 'Managing new product development: some alternative ways to organise the work of technical specialists', *Journal of Marketing Management* 4: 249–60.

Freeman, C. (1974) *The Economics of Industrial Innovation*, Harmondsworth: Penguin.

Fridenson, P. (1978) 'The coming of the assembly line to Europe', *Sociology of the Sciences* 11: 159–75.

Friedman, A.L. (1977) *Industry and Labour: Class Struggle at Work and Monopoly Capitalism*, London: Macmillan.

——— (with Cornford, S.D.) (1989) *Computer Systems Development: History, Organization and Implementation*, Chichester: Wiley & Sons.

Frohman, A. (1985) 'Putting technology into strategy', *Journal of Business Strategy* 5: 54–65.

Gallagher, J.G. and Scott, R.S. (1988) *Kwik-Fit Holdings*, Cranfield: Case Clearing House.

Gallie, D. (1978) *In Search of the New Working Class: Automation and Social Integration within the Capitalist Enterprise*, Cambridge: Cambridge University Press.

Game, A. and Pringle, R. (1983) *Gender at Work*, Sydney: Allen & Unwin.

Gardiner, J.P. (1986) 'Design trajectories for airplanes and automobiles during the past fifty years', in C. Freeman (ed.) *Design, Innovation and Long Cycles in Economic Development*, London: Frances Pinter.

——— and Rothwell, R. (1989) 'Design management strategies', in M. Dodgson (ed.) *Technology Strategy and the Firm: Management and Public Policy*, London: Longman.

Gershuny, J. and Miles, I. (1983) *The New Service Economy*, London: Frances Pinter.

Goold, M. and Campbell A. (1987) 'Managing diversity: strategy and control in diversified British companies', *Long Range Planning* 20: 42–52.

Gouldner, A.W. (1954) *Patterns of Industrial Bureaucracy*, New York: Free Press.

Green, K. (1990) 'Shaping technologies and shaping markets: creating demand for biotechnology', paper presented at the Conference on Firm Strategy and Technical Change, Manchester.

Griffin, R.W. (1983) 'Objective and social sources of information in task design: a field experiment', *Administrative Science Quarterly* 28: 184–200.

—— and Bateman, T.S. (1986) 'Job satisfaction and organisational commitment', in C.L. Cooper and I.T. Robertson (eds) *International Review of Industrial and Organisational Psychology*, Chichester: Wiley.

Grossin, W. (1969) *Le Travail et le Temps: Horaires-durees-rhythmes*, Paris: Editions Anthropos.

Hackman, J.R. and Oldham, G. (1980) *Work Redesign*, Reading, Mass.: Addison-Wesley.

Hampden-Turner, C. (1970) *Radical Man*, New York: Schenkman Publishing.

Harvey-Jones, J. (1988) *Making It Happen: Reflections on Leadership*, Glasgow: Fontana/Collins.

Hassard, J. (1989) 'Time and industrial sociology', in P. Blyton, P. Hassard, S. Hill and K. Starkey, *Time, Work and Organisation*, London: Routledge.

Hedberg, B. and Mumford, E. (1976) 'The design of computer systems', in E. Mumford and H. Sackman (eds) *Human Choice and Computers*, New York: North-Holland Publishing.

Heller, F.A. (1989) 'Human resource management and the socio-technical approach', in G.J. Bamber and R.D. Lansbury (eds) *New Technology*, London: Unwin Hyman.

—— and Hitchon, B. (1979) 'The contribution of employee participation to work design: an example from British Leyland', Conference on Industrial Relations and Working Conditions on the Shop Floor, Hamburg.

Hill, S. (1988) *The Tragedy of Technology*, London: Pluto Press.

Hoskin, K. (1990) 'Using history to understand theory: a reconsideration of the historical genesis of "strategy", paper presented at the EIASM Workshop on Strategy, Accounting and Control, Venice, October.

Howell, J.M. and Higgins, C.A. (1990) 'Champions of technological innovation', *Administrative Science Quarterly* 35: 317–41.

Hughes, T.P. (1987) 'The evolution of large technological systems', in W.E. Bijker, T. Hughes and T.J. Pinch (eds) *The Social Construction of Technological Systems*, Cambridge, Mass. and London: MIT Press.

Hyman, R. (1987) 'Strategy or structure? Capital, labour and control', *Work, Employment and Society* 1: 25–55.

Jordan, N. (1963) 'Allocation of functions between man and machines in automated systems', *Journal of Applied Psychology* 47: 161–5.

Kanter, R.M. (1989) *When Giants Learn to Dance: Mastering the Challenge of Strategy, Management and Careers in the 1990s*, New York: Simon & Schuster.

Keen, P. and Bronsema, G. (1981) *Cognitive Style: A Perspective for Integration*, Working Paper 82. Boston: Massachusetts Institute of Technology, Centre for Information Systems Research.

Keep, E. (1989) 'A training scandal?', in K. Sisson (ed.) *Personnel Management in Britain*, Oxford: Basil Blackwell.

Kelly, J. (1985) 'Management's redesign of work: labour process, labour markets, and product markets', in D. Knights, H. Willmott and D. Collinson (eds) *Job Redesign: Critical Perspectives on the Labour Process*, Farnborough: Gower.

Kern, H. and Schumann, M. (1985) *Das Ende der Arbeitsteilung? Rationalisierung in der Industriellen Produktion*, Munich: Verlag C.H. Beck.

Kesteloot, R. (1989) 'Introduction of computerised numerical control and the rationalisation of the production: the Belgian case', in A. Francis and P. Grootings (eds) *New Technologies and Work: Capitalist and Socialist Perspectives*, London: Routledge.

Kidder, T. (1981) *The Soul of a New Machine*, New York: Avon.

Klein, J.A. (1991) 'An examination of autonomy in the light of new manufacturing practices', *Human Relations* 44: 21–38.

Klein, L. (1978) 'The production engineer's role in industrial relations', *Production Engineer*, December, 27–9.

Kornhauser (1965) *The Mental Health of the Industrial Worker*, New York: Wiley.

Kuhn, T.S. (1962) *The Structure of Scientific Revolutions*, Chicago: University of Chicago Press.

Kulpinska, J. and Skalmierski, A. (1989) 'The taming of new technology: a Polish case study of the introduction of a flexible manufacturing system', in A. Francis and P. Grootings (eds) *New Technologies and Work: Capitalist and Socialist Perspectives*, London: Routledge.

Kusterer, K. (1978) *Workplace Know-How: The Important Working Knowledge of 'Unskilled' Workers*, Boulder, Colorado: Westview Publishing.

Langrish, J., Gibbons, M., Evans, W. and Jevons, F. (1972) *Wealth from Knowledge*, London: Macmillan.

Lash, S. and Urry, J. (1987) *The End of Organized Capitalism*, Cambridge: Polity Press.

Leonard-Barton, D. (1988) 'Implementation as mutual adaptation of technology and organization', *Research Policy* 17: 251–67.

—— (1990) 'The role of process innovation and adaptation in attaining strategic technological capability', paper presented at the Operations Management Association Conference on Manufacturing Strategy: Theory and Practice, Warwick University, June.

Lilienfeld, R. (1978) *The Rise of Systems Theory: An Ideological Analysis*, New York: Wiley.

Littler, C. (1982) *The Development of the Labour Process in Capitalist Societies*, London: Heinemann.

—— (1983) 'A history of new technology', in G. Winch (ed.) *Information Technology in Manufacturing Processes*, London: Rossendale.

Lukes, S. (1974) *Power: A Radical View*, London: Macmillan.

Macdonald, S. (1985) 'Technology beyond machines', in E. Rhodes and D. Wield (eds) *Implementing New Technologies: Choice, Decision and Change in Manufacturing*, Oxford: Basil Blackwell.

McFarlan, F.W. (1984) 'Information technology changes the way you compete', *Harvard Business Review*, May–June, 98–103.

MacKenzie, D. (1990) 'Economic and sociological explanation of technological change', paper presented at the Conference on Firm Strategy and Technical Change, Manchester, September.

────── and Wacjman, J. (eds) (1985) *The Social Shaping of Technology*, Milton Keynes: Open University Press.

McLoughlin, I. (1990) 'Management, work organisation and CAD – towards flexible automation?', *Work, Employment and Society* 4: 217–37.

────── and Clark, J. (1988) *Technological Change at Work*, Milton Keynes: Open University Press.

Majchrzak, A. (1988) *The Human Side of Factory Automation*, San Francisco: Jossey-Bass.

Manwaring, T. (1981) 'The trade union response to new technology', *Industrial Relations Journal* XX.

────── and Wood, S. (1985) 'The ghost in the labour process', in D. Knights, H. Willmott and D. Collinson (eds) *Job Redesign: Critical Perspectives on the Labour Process*, Farnborough: Gower.

March, J.G. and Simon, H.A. (1958) *Organizations*, New York: John Wiley.

────── and Sproull, L.S. (1990) 'Technology, management and competitive advantage', in P.S. Goodman, L.S. Sproull and Associates, *Technology and Organizations*, Oxford: Jossey-Bass.

Marcuse, H. (1968) *Negations*, London: Routledge & Kegan Paul.

Marglin, S.A. (1974) 'The origins and functions of hierarchy in capitalist production', *Review of Radical Political Economy* 6: 60–102.

Markus, M.L. (1984) *Systems in Organisations*, London: Pitman Press.

Martin, C.J. (1986) *Computers and Senior Managers*, Manchester: NCC Publications.

Martin, R. (1988a) 'Technological change and manual work', in D. Gallie (ed.) *Employment in Britain*, Oxford: Basil Blackwell.

────── (1988b) 'The management of industrial relations and new technology', in P. Marginson, P.K. Edwards, R. Martin, J. Purcell and K. Sisson (eds) *Beyond the Workplace: Managing Industrial Relations in the Multi-Establishment Enterprise*, Oxford: Basil Blackwell.

Marx, K. (1954) *Capital* (vol. 1), London: Lawrence & Wishart.

Matthews, J. (1989) *Tools of Change: New Technology and the Democratisation of Work*, Sydney: Pluto Press.

Miller, P. and O'Leary, T. (1987) 'Accounting and the construction of the governable person', *Accounting, Organisations and Society* 12: 58–71.

Mintzberg, H. (1973) *The Nature of Managerial Work*, New York: Harper & Row.

────── (1978) 'Patterns in strategy formulation', *Management Science* 14: 34–49.

────── (1979) *The Structuring of Organizations: A Synthesis of the Research*, Englewood Cliffs, N.J.: Prentice-Hall.

Morgan, G. (1986) *Images of Organizations*, Beverly Hills, Calif.: Sage.

Morgan, G. (1990) *Organisations in Society*, London: Macmillan.

Mueller, W.S., Clegg, C.W., Wall, T.D., Kemp, N.J. and Davies, R.T. (1986) 'Pluralist beliefs about new technology within a manufacturing organisation', *New Technology, Work and Employment* 1: 127–39.

Mumford, E. and Weir, M. (1979) *Computer Systems in Work Design: The ETHICS Method*, London: Associated Business Press.

Mumford, L. (1934) *Technics and Civilization*, New York: Harcourt Brace Jovanovich.

Noble, D.F. (1984) *Forces of Production: A Social History of Industrial Automation*,

New York: Alfred Knopf.

—— (1985) 'Social choice in machine design: the case of automatically controlled machine tools', in D. MacKenzie and J. Wajcman (eds) *The Social Shaping of Technology*, Milton Keynes: Open University Press.

Odaka, K. (1975) *Towards Industrial Democracy: Management and Workers in Modern Japan*, Cambridge, Mass.: Harvard University Press.

O'Reilly, C.A. and Caldwell, D.F. (1985) 'The impact of normative social influence and cohesiveness on task perceptions and attitudes: a social information processing approach', *Journal of Occupational Psychology* 58: 193–206.

Ovitt, G. (1986) 'The cultural context of Western technology: Early Christian attitudes toward manual labour', *Technology and Culture* 27: 477–500.

Pacey, A. (1983) *The Culture of Technology*, Oxford: Basil Blackwell.

Pelto P.J. (1973) *The Snowmobile Revolution: Technology and Social Change in the Arctic*, Menlo Park, Calif.: Cummings.

Perrow, C. (1972) 'Technology, organizations and environment: a cautionary note', paper presented at the Meeting of British Sociological Association (Industrial Sociology group).

—— (1983) 'The organizational context of human factors engineering', *Administrative Science Quarterly* 28: 521–41.

Pettigrew, A.M. (1973) *The Politics of Organisational Decision Making*, London: Tavistock.

—— (1985) *The Awakening Giant: Continuity and Change in ICI*, Oxford: Basil Blackwell.

—— (1987) 'Context and action in the transformation of the firm', *Journal of Management Studies* 24: 649–70.

Pinch, T.J. and Bijker, W.E. (1987) 'The social construction of fact and artifacts: or how the sociology of science and the sociology of technology might benefit each other', in W.E. Bijker, T. Hughes and T.J. Pinch (eds) *The Social Construction of Technological Systems*, Cambridge, Mass. and London: MIT Press.

Piore, M. and Sabel, C. (1984) *The Second Industrial Divide: Possibilities for Prosperity*, New York: Basic Books.

Pollert, A. (1988) 'Dismantling flexibility', *Capital and Class* 34: 42–75.

Porter, M.E. and Millar, V.E. (1985) 'How information gives you competitive advantage', *Harvard Business Review*, July–August, 149–60.

Pralahad, C.K. and Hamel, G. (1990) 'The core competence of the corporation', *Harvard Business Review*, May–June, 79–91.

Rae, J.B. (1965) *The American Automobile: A Brief History*, Chicago: University of Chicago Press.

Ramsay, H. and Beirne, M. (1992) 'Manna or monstrous regiment? Information technology, control and democracy in the workplace', in M. Beirne and H. Ramsay (eds) *Information Technology and Workplace Democracy*, London: Routledge.

Ravetz, J.R. (1971) *Scientific Knowledge and Its Social Problems*, Oxford: Clarendon Press.

Rogers, E.M. (1962) *Diffusion of Innovations*, New York: Free Press.

Rolfe, H. (1990) 'In the name of progress? Skill and attitudes towards technological change', *New Technology, Work and Employment* 5: 107–21.

Rose, M. (1975) *Industrial Behaviour*, Harmondsworth: Penguin.

Rosenberg, N. (1976) *Perspectives on Technology*, Cambridge: Cambridge University Press.

Rothwell, R. (1986) 'Innovation and re-innovation: a role for the user', *Journal of Marketing Management* 2: 109–23.

Rothwell, S. (1984) 'Company employment policies and new technology in manufacturing and service sectors', in M. Warner (ed.) *Microprocessors, Manpower and Society: A Comparative, Cross-National Approach*, Aldershot: Gower.

Rothwell, R., Freeman, C., Horsley, A., Jervis, V.T.P., Robertson, A.B. and Townsend, J. (1974) 'Sappho updated – Project Sappho phase II', *Research Policy* 3: 258–91.

Rubery, J. (1978) 'Structured labour markets, worker organisation and low pay', *Cambridge Journal of Economics* 2: 17–36.

Rubinstein, A.H. (1985) 'Trends in technology management', *IEEE Transactions in Engineering Management*, November, 141–3.

Ruskin College (1985) *Workers and New Technology: Disclosure and Use of Company Information*, Oxford: Ruskin College Press.

Salancik, G. and Pfeffer, J. (1978) 'A social-information processing approach to job attitudes and task design', *Administrative Science Quarterly* 23: 224–53.

Sayer, A. (1986) 'New developments in manufacturing: The Just-In-Time system', *Capital and Class* 30: 43–72.

Scarbrough, H. (1981) 'Working with robots is a bore', *New Scientist*, 28 May, 554–5.

—— (1982) 'The control of technological change in the motor industry: a case study', unpublished Ph.D. dissertation, University of Aston, Birmingham.

—— (1984) 'Maintenance workers and new technology', *Industrial Relations Journal* 15 (4): 9–16.

—— and Lannon, R. (1988) 'The successful exploitation of new technology in banking', *Journal of General Management* 13: 38–51.

—— and Moran, P. (1986) 'Technical change in an industrial relations context', *Employee Relations* 8: 17–22.

Schon, D. (1963) *Displacement of Concepts*, London: Tavistock.

Shaiken, H. (1980) *Computer Technology and the Relations of Power in the Workplace*, discussion paper, Berlin: Institute for Comparative Social Research.

Shimada, H. and MacDuffie, J.P. (1987) *Industrial Relations and 'Human Ware': Japanese Investments in Automobile Manufacturing in the United States*, Cambridge, Mass.: Center for Technology Policy and Industrial Development, MIT.

Shostak, A.B. (1987) 'Technology, air-traffic control, and labor–management relations', in D.B. Cornfield (ed.) *Workers, Managers, and Technological Change*, London: Plenum Press.

Shrivastava, P. and Souder, W.E. (1987) 'The strategic management of technological innovations: a review and a model', *Journal of Management Studies* 24: 25–41.

Silver, J. (1987) 'The ideology of excellence: management and neo-conservatism', *Studies in Political Economy* 24: 105–29.

Singleton, W.T. (ed.) (1978) *The Analysis of Practical Skills*, Lancaster: MTP Press.

Sloan, A.P. (1954) *My Years With General Motors*, New York: Doubleday.

Smith, S. and Wield, D. (1988) 'New technology and bank work: banking on IT as an "organizational technology"' in L. Harris (ed.) *New Perspectives on the Financial System*, Beckenham: Croom Helm.

Sorge, A., Hartmann, G., Warner, M. and Nicholas, I. (1983) *Microelectronics and Manpower in Manufacturing*, Aldershot: Gower.

Stanley, M. (1978) *The Technological Conscience: Survival and Dignity in an Age of Expertise*, New York: Free Press.

Stapp, H. (1972) 'The Copenhagen interpretation and the nature of space–time', *American Journal of Physics* 40: 1098–123.

Stern, N. (1981) *From ENIAC to UNIVAC*, Bedford, Mass.: Digital Press.

Storey, J. (1985) 'The means of management control', *Sociology* 19: 193–211.

Streeck, W. (1989) 'Skills and the limits of neo-liberalism: the enterprise of the future as a place of learning', *Work, Employment and Society* 3: 89–104.

Sviokla, J. (1990) 'An examination of the impacts of an expert system on the firm: the case of XCON', *Management Information System Quarterly*, June, 126–39.

Taggart, W.M. and Silbey, V. (1986) *Information Systems: People and Computers in Organizations*, Boston: Allyn & Bacon.

Taylor, J.C. (1979) 'Job design criteria twenty years later', in L.E. Davis and J.C. Taylor (eds) *Design of Jobs* (2nd edn), Santa Monica, Calif.: Goodyear Books.

Thompson, E.P. (1967) 'Time, work discipline and industrial capitalism', *Past and Present* 38: 56–97.

Thompson, M. and Wildavsky, A. (1986) 'A cultural theory of information bias in organisations', *Journal of Management Studies* 23: 273–86.

Thompson, P. (1989) *The Nature of Work* (2nd edn), London: Macmillan.

——— (1990) 'Crawling from the wreckage: the labour process and the politics of production', in D. Knights and H. Willmott (eds) *Labour Process Theory*, London: Macmillan.

Tierney, M. and Wickham, J. (1989) 'Controlling software labour: professional ideologies and the problem of control', ESRC/PICT Workshop on Critical Perspectives on Software, Manchester, July.

Toffler, A. (1980) *The Third Wave*, London: Collins.

Turner, J.A. and Karasek, R.A. (1984) 'Software ergonomics: effects of computer application design parameters on operator task performance and health', *Ergonomics* 27: 667–74.

Ure, A. (1835) *The Philosophy of Manufactures*, London: Knight.

Veblen, T. (1923) *Absentee Ownership and Business Enterprise in Recent Times*, New York: Charles Scribner.

von Hippel, E. (1976) 'The dominant role of users in the scientific instruments innovation process', *Research Policy* 5: 36–49.

——— (1988) *The Sources of Innovation*, Oxford: Oxford University Press.

——— (1990) *The Impact of 'Sticky Data' on Innovation and Problem-Solving*, Sloan School of Management Working Paper, 3147-90-BPS, MIT, Cambridge, Mass.

Walker, C.R. and Guest, R.H. (1952) *The Man on the Assembly Line*, Cambridge, Mass.: Harvard University Press.

Wallace, M. (1989) 'Brave new workplace: technology and work in the new economy', *Work and Occupations* 16: 363–92.

Weber, M. (1968) *Economy and Society*, New York: Bedminster Press.

Webster, J. (1990) *Office Automation: The Labour Process and Women's Work in Britain*, London: Harvester Wheatsheaf.

Weick, K.E. (1969) *The Social Psychology of Organizing*, Reading, Mass.: Addison-Wesley.

—— (1990) 'Technology as equivoque: sensemaking in new technologies', in P.S. Goodman, L.S. Sproull and Associates, *Technology and Organizations*, Oxford: Jossey-Bass.

Whalley, P. (1986) 'Markets, managers and technical autonomy', *Theory and Society* 15: 223–47.

Whipp, R. and Clark, P.A. (1986) *Innovation and the Auto Industry: Production, Process and Work Organization*, London: Frances Pinter.

White, C. (1981) 'Why won't managers cooperate? Innovation and productivity in engineering', *Industrial Relations Journal* 12: 61–81.

Whittington, R. (1990) 'The changing structures of R&D', in R. Loveridge and M. Pitt (eds) *The Strategic Management of Technological Innovation*, London: Wiley.

—— (1991) 'Changing control strategies in industrial R&D', *R&D Management* 21: 43–53.

Wilkinson, B. (1983) *The Shopfloor Politics of New Technology*, London: Heinemann.

—— and Oliver, N. (1990) 'Obstacles to Japanization: the case of Ford UK', *Employee Relations* 12: 17–21.

Willcocks, L. and Mason, D. (1987) *Computerising work: People, Systems Design and Workplace Relations*, London: Paradigm Press.

Williamson, O.E. (1985) *The Economic Institutions of Capitalism*, New York: Free Press.

Wilson, D.C. (1992) *A Strategy of Change: Concepts and Controversies in the Management of Change*, London: Routledge.

Winch, G. (1983) 'Organisation design for CAD/CAM', in G. Winch (ed.) *Information Technology in Manufacturing Processes*, London: Rossendale.

Winner, L. (1983) 'Technologies as forms of life', in R.S. Cohen and M.W. Wartofsky (eds) *Epistemology, Methodology and the Social Sciences*, Holland: Reidel.

—— (1986) *The Whale and the Reactor*, Chicago: University of Chicago Press.

Womack, J.P., Jones, D.T. and Roos, D. (1990) *The Machine that Changed the World*, Oxford: Maxwell Macmillan International.

Wood, S. (1986) 'The cooperative labour strategy in the US auto industry', *Economic and Industrial Democracy* 7: 415–48.

Woodward, J. (1965) *Industrial Organization: Theory and Practice*, Oxford: Oxford University Press.

—— (ed.) (1970) *Industrial Organization: Behaviour and Control*, Oxford: Oxford University Press.

Woolgar, S. (1985) 'Why not a sociology of machines? The case of sociology and artificial intelligence', *Sociology* 19: 557–72.

Wray, A. (1988) 'The everyday risks of playing safe', *New Scientist*, 8 September, 61–5.

Zimbalist, A. (ed.) (1979) *Case Studies on the Labour Process*, New York: Monthly Review Press.

Zuboff, S. (1988) *In the Age of the Smart Machine*, London: Heinemann.

Zukav, G. (1979) *The Dancing Wu Li Masters: An Overview of the New Physics*, London: Rider/Hutchinson.

Index